"We... About Our Survival, for We Foolishly Jeopardize It..."

"We are wrong to devote our attention to saving or managing nature. Gaia will save herself—with or without us—and hardly needs advice or help in management. To look out for ourselves we would be wise to interfere as little as possible in her ways, and to learn as much as possible of them. Our technology has ravaged nature and continues to do so, but the ravages of technology are based on our unnatural greed, our profit motive. There is no intrinsic reason that we humans cannot develop a benign technology once we agree that our desire to maximize profits is completely at odds with nature's dynamic balance—that greed prevents health and welfare for all. No other creatures take more than they need, and this must be our first lesson...."

"With the artistry of a master storyteller and the enormous erudition of a philosopher-scientist, Elisabet Sahtouris retells the story of creation and evolution...."
—Elinor Gadon,
author of *The Once and Future Goddess*

"Who better than Elisabet Sahtouris can sing the story-song and Gaia-dance so all can hear and... understand?"
—Lynn Margulis,
co-author of *Biospheres*

"Albert Einstein once said that we need a new way of thinking if humanity is to survive. Professor Sahtouris points the way to the kind of thinking that is required. This is a most important book."
—Stephan L. Chorover, professor of psychology,
Massachusetts Institute of Technology

Most Pocket Books are available at special quantity discounts for bulk purchases for sales promotions, premiums or fund raising. Special books or book excerpts can also be created to fit specific needs.

For details write the office of the Vice President of Special Markets, Pocket Books, 1230 Avenue of the Americas, New York, New York 10020.

GAIA
The Human Journey From Chaos To Cosmos

Dr. Elisabet Sahtouris

POCKET BOOKS
New York   London   Toronto   Sydney   Tokyo

An *Original* Publication of POCKET BOOKS

POCKET BOOKS, a division of Simon & Schuster Inc.
1230 Avenue of the Americas, New York, NY 10020

Copyright © 1989 by Dr. Elisabet Sahtouris
Foreword copyright © 1989 by J. E. Lovelock
Cover art copyright © 1989 Jean-Francois Podevin

All rights reserved, including the right to reproduce this book or portions thereof in any form whatsoever. For information address Pocket Books, 1230 Avenue of the Americas, New York, NY 10020

ISBN: 0-671-68002-1

First Pocket Books trade paperback printing May 1989

10 9 8 7 6 5 4 3 2 1

POCKET and colophon are trademarks of Simon & Schuster Inc.

Printed in the U.S.A.

To my planet and its people

Contents

	A NOTE FROM THE AUTHOR	9
	FOREWORD BY JAMES E. LOVELOCK	13
1.	A TWICE-TOLD TALE	19
2.	COSMIC BEGINNINGS	29
3.	THE YOUNG EARTH	38
4.	GAIAN PROBLEMS	50
5.	THE DANCE OF LIFE	63
6.	A GREAT LEAP	76
7.	EVIDENCE OF EVOLUTION	87
8.	FROM PROTISTS TO POLYPS	100
9.	FROM POLYPS TO POSSUMS	111
10.	FROM POSSUMS TO PEOPLE	122

11.	THE BIG BRAIN EXPERIMENT	133
12.	WHAT THE PLAY IS ALL ABOUT	146
13.	WORLDVIEWS FROM THE PLEISTOCENE TO PLATO	158
14.	WORLDVIEWS FROM PLATO TO THE PRESENT	170
15.	LESS THAN PERFECT, MORE THAN MACHINE	181
16.	THE BODY OF HUMANITY	192
17.	A MATTER OF MATURATION	205
18.	ECOLOGICAL ETHICS	218
19.	COSMIC CONTINUATION	230
	EPILOGUE	242
	BIBLIOGRAPHY	247

A Note from the Author

This book is a work of philosophy in the original sense of a search for wisdom, for practical guidance in human affairs through understanding the natural order of the cosmos to which we belong. It bears little resemblance to what we have come to call philosophy since that effort was separated from natural science and became more an intellectual exercise than a practical guide for living.

To find meaning and guidance in nature, I integrated my personal experience of it with those scientific accounts that seemed to best fit it. From this synthesis, meaning and lessons for humanity emerged freely. I did the work in the peaceful, natural setting of a tiny old village on a small pine-forested Greek island, where I could consider the research and debates of scientists, historians, and philosophers, then test them against the natural world I was trying to understand.

Putting into simple words the specialized technical language of scientists and winding my way through labyrinths of philosophic prose, I gradually simplified the story of the origins and nature of our planet within the larger cosmos, and of our human origins, nature, and history within the larger being of this planet.

The Gaia hypothesis, now Gaia theory, of James Lovelock and Lynn Margulis—the theory that our planet and its creatures constitute a single self-regulating system that is in fact a great living being, or organism—is the conception of physical reality in which my philosophy is rooted. Quite simply, it makes more sense on all levels—intuitive, experiential, scientific, philosophical, and even aesthetic and ethical—than any other conception I know. And I have come to believe, in the course of this work, that this conception contains profound and pressing implications for all humanity.

To ensure that my vision of evolution and history would stay simple and in clear focus, I kept telling its essence and more than a few of its particulars in something of the style of an ancient storyteller (albeit less poetically) during many social evenings among my Greek village friends. I also wrote the story in English for children before I set about an adult version.

To my surprise, these deliberate exercises in simplicity proved more difficult than writing for professional audiences, for in stripping our intellectual language to the essence of what is being said, we must be very sure that essence is really there, really coherent. Science has been a process of differentiating our knowledge into an incredible wealth of precise details, but these details become ever more disconnected from one another and cry out for integration into coherent wholes. I have no doubt I will be accused of oversimplification, and perhaps rightly so, as one pays for scope in lack of detail and precision.

Friends and colleagues have asked me now and then why I insist on dealing with all evolution, even all the cosmos, to discuss human matters, why I don't narrow my scope to workable proportions. My answer is that context is what gives meaning, and a serious search of context is an ever-expanding process leading inevitably to the grandest context of all: the whole cosmos. As the nested contexts for the human story—especially the context of evolution—became clearer to me, they revealed a simple but elegant biological vision of just why our human condition has become so critical and what we might do to improve it.

Other people ask why I'm so eager to save humanity when it is proving such a social and ecological disaster. To this I can only answer that, as far as I can see, every healthy living being or system in nature has evolved behavior consistent with its survival, and I do not exclude myself from this natural health scheme.

I can no more proclaim the worldview arising from my work "reality" than can any particular philosopher working at creating a meaningful worldview in any particular place and time, drawing on the scientific and historical knowledge of that place and time. Philosophy is an intensely personal search that one hopes will have relevance to others, will be validated by their experience, will offer them some insight and guidance, or will at least stimulate them in their disagreement to search further on their own.

Yet a work of philosophy also reflects the broader context and search of a culture at a particular stage, and the biological evolutionary viewpoint of this book reflects a broadly emerging pattern of search for our origins and direction in nature—a reawakening of that search begun by the original pre-Socratic philosophers, indeed that goes further back to the roots of religion—the search for *re-ligio*, for "reconnection" with our origins in nature or cosmos, within which we were created and within which we continue our creation.

Paradoxically, our self-imposed separation from nature by way of an "objective" mechanical worldview during the past few millennia has led to the scientific knowledge that makes it possible to understand and reintegrate ourselves into nature's self-organization. It has also brought us to a stage of technology that permits us to share our discoveries and our understanding planet-wide in no time at all, to work together as a body of humanity with hope of transcending our present crisis in a far healthier and happier future for ourselves and all the rest of Earthlife.

Although the actual work of this book was done in relative isolation and without funding, I am indebted and profoundly grateful to many teachers and friends, from the forest creatures with whom I spent my earliest years to Jim Lovelock and Lynn Margulis, who have not only informed and inspired me in these recent years, but who have given me invaluable encouragement, confidence, and opportunities in seeing the work through.

ELISABET SAHTOURIS
Metochi, Agistri, Greece
July 1988

Foreword

O The Gaia hypothesis, now accorded the status of Gaia theory, is maturing with experience and the tests of time, not unlike the humans of this book. It is spurring a great deal of scientific research into the geophysiology of our living planet. It is also spurring philosophic conceptions of what it means to our species to be part of a living planet. Some of these conceptions stay carefully within the accepted limits of science; others have a religious bent. Most, especially environmentalist conceptions, advocate for humanity, being primarily concerned with human survival. A few, taking a clue from my partner Lynn Margulis and myself, advocate for the planet and the much maligned microbes with which the Gaian system originated and which continue to do its basic work.

Elisabet Sahtouris's conception integrates scientific Gaian evolution with the human search to connect with our roots, inspiring us to learn from billions of years of Gaian experience in the self-organization of workable living systems. It is well balanced between advocacy for the planet and advocacy for humans, placing the onus on humans to recognize the lack of maturity involved in believing we can manage the planet, and to learn instead to follow its lead in organizing ourselves.

Elisabet gives us valuable insights as she draws parallels between the evolution of cells and the evolution of human society, pointing out the contrast between the healthy organization of cells, bodies, and biosystems on the one hand and the unhealthy organization of economics and politics in human society on the other. While she argues that our social evolution is not as much under our control as we like to think, she warns us that our survival depends on our meeting the evolutionary demand to transform competitive exploitation into cooperative synergy.

On the whole, her advice makes sense because she herself has taken the trouble to learn directly from nature as well as from the growing store of scientific knowledge about nature. I began the preface to my own book *The Ages of Gaia* by saying that the place in which it was written was relevant to its understanding. Living and working in the Devonshire countryside, far from universities and large research organizations, makes me an eccentric as a scientist, but, as I said, it is the only way to work on an unconventional topic such as Gaia. When I met Elisabet, having accepted her invitation to trace Gaia's roots in Greece, I recognized her as a kindred spirit. She had abandoned academia for a simple lifestyle in the kind of natural setting that brings one closer to understanding what our planet and our species are all about; she was free to develop her own conception of Gaia through a synthesis of scientific knowledge and personal experience of nature. To my surprise, she expressed some concern, some guilt, at having abandoned her profession of science for a pleasant existence in a forest overlooking the sea, the kind of forest that had been home to her in childhood, where she could work out the meaning of things for herself. As I read her work in progress, I was able to assure her she could never have done anything comparable in a constrained academic setting.

In the intervening years, even in the short time since I wrote my own words about Gaia being an unconventional topic, less eccentric scientists than I have declared Gaia more conventional, meaning that Gaia theory is now recognized as a legitimate and fruitful basis for scientific investigation and is thus being brought into the scientific fold. In our first account of Gaia as a system, neither Lynn Margulis nor I fully understood what it was we were describing. Our language tended to be anthropomorphic and, especially in my first book, *Gaia,* poetic. Not surprisingly, some scientists misunderstood our intentions and accused us of saying that organisms acted from some in-built purpose to regulate the planet's climate and chemical composition. The notion of purpose in natural systems is of course a scientific taboo, a sin of heresy. That heresy is avoided in the clearer modern version, which is Gaia theory. This theory sees the evolution of the material environment and the evolution of organisms as tightly coupled into a single and indivisible process or domain. Gaia, with its capacity for homeostasis, is an emergent property of this domain. There is no more need to invoke notions of purpose or foresight in the evolution of this domain than there is in the evolution of our own bodies within Gaia.

As the title of a recent article in *Science* put it, "No Longer Willful, Gaia Becomes Respectable." This means that Gaia scientists will be constrained by bureaucratic forces, by the pressures of tenure, and by the tribal divisions and rules of scientific disciplines. That, in turn, means we will need some antidote to the inevitable separations and constraints. We will need independent synthesizers and visionaries who can make sense of the data produced by the scientific establishment and present it to us in ways that make our living planet real to us within the Gaian context and thus give meaning to our own lives and those of our children and grandchildren.

This is what Elisabet Sahtouris's work means to me, for she comfortably integrates the traditionally separated domains of biology, geology, and atmospheric science to show us the evolution of our living planet and our own roots within it. She then inspires us on ethical grounds to learn from this planetary organism of which we are part, showing us how we can mature as a species well integrated into the larger dance of life.

Elisabet uses the metaphor of dance effectively for its concepts of improvisation and evolution, the creation of order from chaos, the myriad patterns that can be created from a few basic steps. I am myself an inventor of scientific instruments, and so it is second nature to me to think in terms of mechanical and mathematical models. Cybernetic models have proved especially useful in my work of demonstrating how Gaian homeostasis, such as maintaining the earth's temperature, might work. Yet I quite agree with Elisabet that any model we make of nature is at heart metaphorical in that it begins with some image or formula familiar to us humans and used to represent the complexities of nature in simple, understandable, and useful ways. No metaphor should be mistaken for reality, and perhaps a variety of metaphors is insurance against the temptation to do so. I am increasingly impressed by scientists and philosophers who find nonmechanical metaphors for natural systems useful in interpreting Gaia theory.

Elisabet's analysis of science reflects a trend that may well make science in the near future as unrecognizable as today's science would be to the ancients. She does well to remind us that science is a human activity that evolves, a living system in which conservatism should be balanced by healthy controversy. After all, as she so well describes, all Gaian systems are forever busy working out their cooperation through conflicting interests, their unities through diversity.

The optimistic view this book radiates, that despite our errors and immaturities we can still become a healthy species within a healthy planet, is much needed in this age of doomsday predictions. Though time is growing short in our continued destruction of forests, atmospheres, and other critical Gaian systems, nothing would make me happier personally than to see Gaia theory useful in bringing about a better world for Gaia and her people.

JAMES E. LOVELOCK
Coombe Mill, Cornwall, England
July 1988

GAIA

A Twice-Told Tale

Everyone knows that humanity is in crisis—politically, economically, spiritually, ecologically—any way you look at it. Many see humanity as close to suicide by way of our own technology; many others see humans as deserving God's or nature's wrath in retribution for our sins. However we see it, we are deeply afraid that we may not survive much longer. Yet our urge to survival is the strongest urge we have, and we do not cease our search for solutions in the midst of crisis.

The proposal made in this book is that we see ourselves in the context of our planet's biological evolution, as a still new, experimental species with developmental stages that parallel the stages of our individual development. From this perspective, humanity is now in adolescent crisis and, just because of that, stands on the brink of maturity—in a position to achieve true humanity in the full meaning of that word. Like an adolescent in trouble, we have tended to let our focus on the crisis itself or on our frantic search for particular political, economic, scientific, or spiritual solutions depress us and blind us to the larger picture, to avenues of real assistance. If we humbly seek help instead from the nature that spawned us, we will find biological clues to solving all our biggest problems at once. We will see how to make the healthy transition into maturity.

Some of these biological clues are with us daily, all our lives, in our own bodies; others can be found in various ages and stages of the larger living organism of which we are part—planet Earth. Once we see these clues, we will wonder how we could have failed to find them for so long.

The reason we *have* missed them is that we have not understood ourselves as living beings within a larger being, in the same sense that our cells are part of each of us. Our intellectual heritage for thousands of years, most strongly developed in the past few hundred years of science, has been to see ourselves as separate from the rest of nature, to convince ourselves we see *it* objectively—at a distance from ourselves—and to perceive or at least model it as a vast mechanism.

This objective mechanical worldview was founded in ancient Greece when philosophers divided into two schools of thought about the world—one that all nature, including humans, was alive and *self*-creative, ever making order from disorder; the other that the "real" world could be known only through pure reason, not through direct experience, and was God's geometric creation—permanently mechanical and perfect behind our illusion of its disorder. This mechanical/religious worldview superseded the older one of living nature to become the foundation of the whole Western worldview up to the present.

Philosophers such as Pythagoras, Parmenides, and Plato were thus the founding fathers of our mechanical worldview, though Galileo, Descartes, and other men of the Renaissance translated it into the scientific and technological enterprise that has dominated human experience ever since.

What if things had gone the other way? What if Thales, Anaximander, and Heraclitus, the organic philosophers who saw all the cosmos as alive, had won the day back in that ancient Greek debate? What if Galileo, as he experimented with both telescope and microscope, had used the latter to seek evidence for Anaximander's theory of biological evolution here on earth, rather than looking to the skies for confirmation of Aristarchus's celestial mechanics? In other words, what if modern science and our view of human society had evolved from organic biology rather than from mechanical physics?

We will never know how the course of human events would have differed had they taken this path, had physics developed in the shadow of biology rather than the other way around. Yet it seems we were destined to find the biological path eventually, as the mechanical worldview we have lived with so long is now giving way to an organic view—in all fairness, an organic view made possible by the very technology born of our mechanical view.

The same technology that permits us to reach out into space has permitted us to begin seeing the real nature of our own planet—to discover that it is alive and that it is the only live planet circling our sun. The implications of this discovery are enormous, and we have hardly even begun to pursue them. We were awed by astronauts' reports that the earth looked from space like a living being, and were ourselves struck by its apparently live beauty when the visual images were before our eyes. But it has taken time to accumulate scientific evidence that the earth is a live planet rather than a planet with life upon it, and many scientists continue to resist the new conception because of its profound implications for change in all branches of science, not to mention all society.

The difference between a planet with life on it and a living planet is hard at first to understand. Take for example the word, the concept, the practice of "ecology," which has become familiar to us all within just the few short decades that we have been aware of our pollution and destruction of the environment on which our own lives depend. Our ecological understanding and practice has been a big, important step in understanding our relationship to our environment and to other species. Yet, even in our serious environmental concern, we still fall short of recognizing ourselves as part of a much larger living organism. It is one thing to be careful with our environment so it will last and remain benign; it is quite another to know deeply that our environment, like ourselves, is part of the body of an earth organism.

Despite the fact that our study of the earth as a live organism is still new, we already know something of its development as an "embryo" and of its present physiology. The earth transformed itself into a wealth of living species organized into environments in more or less the same way that fertile eggs transform themselves into a wealth of different cells organized into organs, except that in earth's case the organism matures without breaking its shell.

The earliest species into which the materials of the earth's crust transformed themselves created their own environments, and these environments in turn shaped the fate of species, much as cells create their surround and are created by it in our own embryological development. As for physiology, we already know that the earth regulates its temperature—as well as any of its warm-blooded creatures—such that it stays within bounds that are healthy for life despite the sun's steadily increasing heat. And just as our bodies continually renew and adjust the balance of chemicals in our skin and blood, our bones and other tissues, so does the earth continually renew and adjust the balance of chemicals in its atmosphere, seas, and soils. How these physiological systems work is now partly known, partly still to be discovered, as is also still the case with our bodies' physiological systems. Certainly it is ever more obvious that we are not studying the mechanical nature of Spaceship Earth but the self-creative, self-maintaining physiology of a live planet.

We still take the newer Live Earth concept—named Gaia after the earth goddess of early Greek myth—more as a poetic or spiritual metaphor than as a scientific reality. However, the name Gaia was never intended to suggest that the earth is a female being—the reincarnation of the Great Goddess or Mother Nature herself—or to start a new religion (though it would hardly hurt us to worship our planet as the greater Being whose existence we have intuited from time immemorial). It was intended simply to designate the concept of a live earth, in contrast to an earth with life upon it.

Actually, "Gaia," or the Roman form, "Gea," was an earlier name for our planet than "Earth." It was lost in the wandering of words from ancient Greek through other languages to English. In Greek, our planet has always been called Gaia in its alternate spelling "Ge," which did get into the English language words "geology," the formation of the earth; "geometry," the measurement of the earth; and "geography," the mapping of the earth. In accord with our own practice of calling planets by the names of Greek deities in their Roman versions, we really should call the earth "Gea." Greek, like English, has always used the same word for earth-as-world and earth-as-ground—the ancient "Ge" that became the modern "Gi," pronounced "Yee." The English word "earth" came from an ancient Greek root meaning "working the ground," or earth—*ergaze*—which evolved into the name of the Nordic earth goddess, Erda, and then into the German Erde and the English Earth. Thus even the word "earth" implies a female deity.

With that digression intended to make the name Gaia more acceptable to those who still consider the name and image somehow inappropriate for a scientific concept, let us look also at the myth itself—the creation myth of Gaia's dance.

○ ○ ○

The story of Gaia's dance begins with an image of swirling mist in the black nothingness called Chaos by the ancient Greeks—an image reminding us of modern photos of galaxies swirling in space. In the myth it is the dancing goddess Gaia, swathed in white veils as she whirls through the darkness. As she becomes visible and her dance grows ever more lively, her body forms itself into mountains and valleys; then sweat pours from her to pool into seas, and finally her flying arms stir up a wind-sky she calls Ouranos—still the Greek word for sky—which she wraps around herself as protector and mate.

Though she later banishes Ouranos—Uranus, in Latin—to her depths for claiming credit for creation, their fertile union as Earth and Heaven brings forth forests and creatures including the giant Titans in human form, who in turn give rise to the gods and goddesses and finally to mortal humans.

From the start, says the myth—true to human psychology—people were curious to know how all this had happened and what the future would bring. To satisfy their curiosity, Gaia let her knowledge and wisdom leak from cracks in the earth at places such as Delphi where her priestesses interpreted it for people.

Our curiosity is still with us thousands of years after this myth served as explanation of the world's creation. And in a sense, Gaia's knowledge and wisdom are still leaking from her body—not just at Delphi, but everywhere we care to look in a scientific study of our living planet.

The new scientific story of Gaian creation has other parallels to the ancient myth. We now recognize the earth as a single self-creating being that came alive in its whirling dance through space, its crust transforming itself into mountains and valleys, the hot moisture pouring from its body to form seas. As its crust became ever more lively with bacteria, it created its own atmosphere, and the advent of sexual partnership finally did produce the larger life forms—the trees and animals and people.

The tale of Gaia's dance is thus being retold as we piece together the scientific details of our planet's dance of life. And in its context, the evolution of our own species takes on new meaning in relation to the whole. Once we truly grasp the scientific reality of the Gaian organism and its physiology, our entire worldview and practice are bound to change profoundly, revealing the way to solving what now appear to be our greatest and most insoluble problems.

○ ○ ○

From a Gaian point of view, we humans are an experiment—a trial species still at odds with ourselves and other species, still not having learned to balance our own dance within that of our whole planet. Unlike most other species, we are not biologically programmed to know what to do; rather, we are an experiment in free choice. This leaves us with enormous potential, powerful egotism, and tremendous anxiety—a syndrome that is recognizably adolescent.

Human history may seem very long to us as we study all that has happened in it, but we know only a few thousand years of it and have existed as humans for only a few million years, while Gaian creation has been going on for billions of years. We have scarcely had time to come out of species childhood, yet our social evolution has changed us so fast that we have leaped into our adolescence.

Humans are not the first Gaian creatures to make problems for themselves and for the whole Gaian system, as we will see. We are, however—unless whales and dolphins beat us to it in past ages—the first Gaian creatures who can understand such problems, think about them, and solve them by free choice. In fact, the argument of this book is that our maturity as a species depends on our accepting the responsibility for our natural heritage of behavioral freedom by working consciously and cooperatively toward our own health along with that of our planet.

Our ability to be objective, to see ourselves as the "I" or "eye" of our cosmos, as beings independent of nature, has inflated our egos—"ego" being the Greek word for "I." We came to separate the "I" from the "it" and to believe that "it"—the world "out there"—was ours to do with as we pleased, telling ourselves we were either God's favored children or the smartest and most powerful naturally evolved creatures on earth. This egotistic attitude has been very much a factor in bringing us to adolescent crisis. And so an attitude of greater humility and willingness to accept some guidance from our parent planet will be an important factor in reaching our species maturity.

The tremendous problems confronting us now—inequality, hunger, the threat of nuclear annihilation and possibly irreversible damage to the natural world we depend on as much as any of our cells depend on the wholeness of our bodies for their life—are all of our own making. These problems have become so enormous that many of us believe we will never be able to solve them. Yet just at this time in our troubled world we stand on the brink of maturity, in a position to recognize that we are neither perfect nor omnipotent, but that we can learn a great deal from a parent planet that is also not perfect or omnipotent but has the experience of billions of years of overcoming an endless array of difficulties, small and great.

When we look anew at evolution, we see not only that other species have been as troublesome as ours, but that many a fiercely competitive situation resolved itself in a cooperative scheme. The kind of cells our bodies are made of, for example, began with the same kind of exploitation among bacteria that characterizes our historic human imperialism. And through the same technologies of transportation and communications first invented by those bacteria as they bound themselves into the cooperative venture that made our existence possible, we are uniting our selves into a single body of humanity that may make yet another new step in Gaian evolution possible. If we look to the lessons of evolution, we will gain hope that the newly forming worldwide body of humanity may also learn to adopt cooperation in favor of competition. The necessary systems have already been invented and developed; we lack only the understanding, motive, and will.

It may come as a surprise that nature has something to teach us about cooperative economics and politics. Sociobiologists who have told us much in recent decades about humanity's animal heritage have tended to paint us a bleak picture—calling on our evolutionary heritage as evidence that we will never cure ourselves of territorial lust and aggression toward one another, and that thus there will be no end to economic greed and political warfare. But it is the aim of this book to show that these sociobiologists have presented a misleading picture—as misleading as earlier scientists' one-sided view of all natural evolution as "red in tooth and claw," the hard and competitive struggle among individuals on which we have modeled our modern societies.

The new view of our Gaian earth in evolution shows, on the contrary, an intricate web of cooperative mutual dependency, the evolution of one scheme after another that harmonizes conflicting interests. The patterns of evolution show us the creative maintenance of life in all its complexity. Indeed nature is more suggestive of a mother juggling resources to ensure each family member's welfare as she works out differences of interest to make the whole family a cooperative venture, than of a rational engineer designing perfect machinery that obeys unchangeable laws. For scientists who shudder at such *anthropomorphism*—giving nature human form—let us not forget that *mechanomorphism*—giving nature mechanical form—is really no better than secondhand anthropomorphism, since mechanisms are human products. Is it not more likely that nature in essence resembles one of its own creatures than that it resembles in essence the nonliving product of one of its creatures?

The leading philosophers of our day recognize that the very foundations of our knowledge are quaking—that our understanding of nature as machinery can no longer be upheld. But those who cling to the old understanding seriously fear that all human life will break down without a firm foundation for our knowledge of nature in mathematical reference points and laws of physics. They fail to see what every child can see—that hummingbirds and flowers *work,* that nature does very well in utter ignorance of human conceptions of how it must work.

Machinery is in fact the very antithesis of life. One must always hope a machine, between its times of use, will not change, for only if it does not change will it continue to be of use. Left to its own devices, it will eventually be destroyed by its environment. Living organisms, on the other hand, cannot stay the same *without* changing constantly, and they use their environment to their advantage. To be sure, our machinery is getting better and better at imitating life; if this were not so, a mechanical science could not have advanced in understanding. But mechanical models of life continue to miss its essential self-creativity.

○ ○ ○

We are learning that there is more than one way to organize working systems, to produce order and balance; that the imperfect and flexible principles of nature lead to greater stability and resilience in natural systems than we have produced in ours—both technological and social—by following the mechanical laws we suppose to be natural.

We design our societies as though they were machinery; we make war on one another over who has the perfect social design. Our greatest conflict is over whether individuals should sacrifice their individual interest to the welfare of the whole or whether individual interest should reign supreme in the hope that the interests of the whole will thus take care of themselves.

No being in nature, outside our own species, is ever confronted with such a choice, and if we consult nature, the reason is obvious. The choice makes no sense, for neither alternative can work. No being in nature can ever be completely independent, although independence calls to every living being, whether it is a cell, a creature, a society, a species, or a whole ecosystem. Every being is part of some larger being, and as such its self-interest must be tempered by the interests of the larger being to which it belongs. Thus mutual consistency works itself out everywhere in nature, as we will see again and again in this book. For clues on organizing a workable economics and politics, we need not even look beyond our own bodies, with their cooperative diversity of cells and organs as a splendid example to us in working out our social future.

Diversity is crucial to nature, yet we humans seem desperately eager to eliminate it, in nature and in one another. This is one of the greatest mistakes we are making. We reduce complex ecosystems to one-crop "economies," and we do everything in our power to persuade or force others to adopt our languages, our customs, our social structures, instead of respecting theirs. Both practices impoverish and weaken us within the Gaian system.

We are right to worry about our survival, for we foolishly jeopardize it. We are wrong to devote our attention to saving or managing nature. Gaia will save herself—with or without us—and hardly needs advice or help in management. To look out for ourselves we would be wise to interfere as little as possible in her ways, and to learn as much as possible of them. Our technology has ravaged nature and continues to do so, but the ravages of technology are based on our unnatural greed, our profit motive. There is no intrinsic reason that we humans cannot develop a benign technology once we agree that our desire to maximize profits is completely at odds with nature's dynamic balance—that greed prevents health and welfare for all. No other creatures take more than they need, and this must be our first lesson. Our second is to learn and emulate nature's fine-tuned recycling system, largely powered by free solar energy.

The purpose of this book is to help pave the way to a happier and healthier future through an understanding of our relationship to the Gaian system that spawned us and of which we are part—a great being that, however it may annoy us, is not ours to dominate and control. We can damage it, but we cannot run it; we had better try to find out what it is all about and what we are doing, and may do, within it.

The aggressive and destructive motives of domination, conquest, control, and profit have been presented to us as human nature by historians as well as by sociologists. But mounting evidence from archaeology strongly suggests that human societies were for the greater part of civilized history based on cooperation and reverence for life and nature, not on competition and obsession with death and technology. It seems our human childhood—which lasted far longer than has our recent adolescence—was guided by religious images of a near and nurturing Mother Goddess before a cruel and distant Father God replaced her in influence. As we come out of adolescence we often recognize the value of what we were taught in childhood, and this new historical view of ourselves supports the general thesis of this book.

Like Gaian creation itself, human understanding or knowledge ever evolves. Parts of the story you are about to read will already have changed by the time you read it. Others will change in the years to come as new things about Earth-Gaia and about human history are discovered. Any of us is free to help find new pieces of the story, bring those we know up to date, and then reinterpret the evidence as a whole, for in the last analysis, every interpretation has its personal color and flavor.

The next chapter is concerned with cosmic beginnings as a living context for our living planet; succeeding chapters, up to half of this book, tell of Gaian evolution over billions of years before we humans become part of it. Those interested in the story of human society may be tempted to skip this part of the story, but the scientific account of evolution in this book is not separable from our human social history. The details of our biological heritage from ancient bacteria on are given because therein lie the clues to a better human future. It is only within this context that we can appreciate our newness and our differences from the rest of nature, to see at the same time how we can benefit from its vast experience to fit ourselves in more harmoniously.

It is on this that everything now depends; species suicide is our only alternative, and there is really no reason to make a dramatic adolescent exit instead of growing up, taking on adult responsibility, and reaping the pleasures of productive maturity. Let us then follow the evolution of Gaian creation and of our own history as social and technological creatures within this great dance of life. Let's see what meaning and guidance all this may give in our present crisis, to speed us on our way into full maturity—to a happier future in which we promote our own health and that of our planet within the greater cosmic dance.

Cosmic Beginnings

The Greek myth of Gaian creation began with an image of the goddess whirling out of darkness, wrapped in flowing white veils. In ancient India the very beginning of the universe, or cosmos, was imagined as swirls in a sea of milk.

We will probably never know how ancient peoples understood that the first forms to create themselves were whirling white spirals. However they knew, we in our own day can actually see just what those first swirling white forms out in space really were. We call them *protogalaxies,* or first galaxies. And we have learned that whole protogalaxies do dance as whirling white forms in space long before planets evolve within them, and longer before creatures can evolve as parts of planets.

The universe, as scientists can best describe it today, began with a huge explosion of energy they call the Big Bang. Some say this explosion was more like a great wave of energy rising out of an even greater sea of energy. Whatever happened to start our universe, we believe with reasonable confidence that it began as very hot, explosively fast-moving energy that has been spreading and cooling ever since, creating space-time as it does so.

The ancient Greek word *chaos,* remember, first denoted the nothingness before there was anything in the universe. Later it came to mean anything so mixed up or messed up that it has no pattern, no order, no meaning—at least none that we humans can detect. (The word "random" carries the same meaning of lack of order or pattern.) Both these ways of using the word "chaos" describe the beginning of the universe quite well. There was nothingness, as no-thing had been formed. Yet the dance of energy that would create order or pattern had begun.

The word *cosmos* was coined as the opposite of *chaos*—to mean order as opposed to disorder, form and pattern instead of formlessness and patternlessness, things instead of no-things, a world instead of no world. The first Greek philosophers understood creation as a process of turning disorderly chaos into an orderly cosmos, and we have no better way of describing it today. For as the chaotic hot energy cooled and spread, it turned itself into a great dance of spiraling cosmic patterns.

Our best explanation of how this happened begins with the idea of imbalance, as it also did in many ancient philosophies. In the early chaos, as the energy of the Big Bang spread and cooled, there must have been pockets of more or less energy—or, as energy formed itself into particles, pockets of more or fewer particles, or different numbers of different kinds of particles. Any such imbalances would have set up currents of motion among the heavier slower-moving particles in the overall force of out-thrusting universal energy.

Particles, or subatomic particles, are the tiniest whirling packets of pure energy from which all matter—all the stuff of the universe—is made. The whirling energy of particles created a new force, or forces, among particles, so that when early cosmic particles passed close enough to each other to attract each other, some of them held together as simple atoms. We can imagine this as rather like people dancing, attracting each other when close enough to whirl each other about. Other particles were pushed apart, while most particles kept zooming along alone among the first slower atoms of floating gas.

The physical force that still works at the greatest distance among the clumps of matter that formed in our universe is the one we call gravitation; two others—the strong and weak nuclear forces—have their effect inside atoms and stars. The fourth and last to develop was the electrical force, which works to combine atoms into molecules—but that is getting ahead of our story.

Natural, or physical, forces, then, on great and small levels, pulled and pushed the universe into patterns great and small. As the number of atoms, and the explosive young universe itself, grew larger, imbalances here and there drifted and swirled the atoms into great gas clouds. These clouds formed more swirls within themselves, some of the thickest becoming protogalaxies sparking with light.

Light is made of energy packets we call *photons*. New photons can be created like tiny sparks when other fast-flying particles bump into one another very hard. Photons make the protogalaxies visible.

If an ancient storyteller could have looked through a modern telescope to see a protogalaxy forming, he might well have said, "Ah, you see, there is the white-veiled Gaia whirling about in her dance." A modern scientist, on the other hand, sees such protogalaxies as the natural result of imbalances and forces in the great cosmic energy field—a swirling of disorderly or chaotic matter into orderly or cosmic patterns; a sea of energy whose forceful currents form natural whirlpools large and small. This is especially important to recognize: that the largest patterns—the great swirling clouds within which protogalaxies took shape—were forming almost as soon as the tiniest particles and atoms began whirling into being. Our universe, or cosmos, has always been a dance of interactions among the large and small moving patterns, each contributing to the other's formation.

But can we really see protogalaxies forming billions of years ago while looking through telescopes now? Is it possible to look back into time, and so very far back at that?

We can. With modern telescopes we can see back to nearly the beginning of the universe! Magical as it seems, the explanation for this strange power we have is quite simple. Everything we see comes to our eyes as light photons that have bounced off or come out of whatever we are looking at. Light bounces off a cat or a cloud, for instance, and comes out of a candle flame or a star. But what exactly is light?

We've already talked about photons as energy particles created when other particles bump into one another. Stars and flames are made of atoms and particles moving so fast that many photons are created in them.

Photons travel through space in waves of different lengths and strengths, some of which we see as different colors and brightnesses when they get to our eyes. Though light is the fastest energy we know, it still takes some time to get from an object that created it, or from one it has bounced off, to our eyes. The time it takes light to travel holds the secret of looking back in time.

It takes about seven minutes for light to get from the sun to our eyes. Every time we look at the sun, we are seeing the light pattern that left it seven minutes ago. That means we are seeing the sun the way it *was* seven minutes ago and not as it is the moment we are looking at it. The sun is the star nearest to us. Other stars are so far away that their light takes years to get to our eyes—thousands of years, even millions of years, depending on how far away the star is. The distance of stars, in fact, is measured in light-years—the number of years it takes for their light to reach us.

Whenever you look up at the night sky, even without a telescope, you are looking back into time. You see each star as it was when the light reaching your eyes left it. By looking at many stars, you are looking at many times past. How far past depends, of course, on the distance of each star. The farther away the star is, the longer ago it sent out the information about what it looks like—that is, the light pattern of the star that has finally found your eyes.

Our own galaxy, the Milky Way, is shaped like a giant swirling pinwheel. It takes light a hundred thousand years to cross it. If there are any creatures on another planet—say, three thousand light-years from us, in our own galaxy—who are looking at us right now, what do they see? If their telescopes are powerful enough, they may be seeing a storyteller speaking of Gaia's dance in an ancient Greek village!

Powerful telescopes can pick up light that is too weak from its long travels for our eyes alone to see—even light from stars and galaxies so old that they were among the first stars and galaxies, or protogalaxies, in the universe, so old they are just beginning cosmic creation. Let's watch one of them in its evolving dance.

Inside the spiraling veils of hydrogen gas, which is made of the first and simplest atoms in the universe, smaller rolling waves create a ring of denser atoms, of more intense energy, at the center. Around it, great loose balls of gas form, something like the way dust balls form under a bed. In the center of such balls, the lively atoms and particles are pulled ever closer together by physical forces until it gets very hot from all the crowding. As these gas balls get hotter and heavier, they become stars.

Wherever we look in the ancient skies, we see galaxies taking shape and growing through different stages. Inside the first generation of stars the incredible heat and pressure begins causing what we now know as *nuclear reactions*—the transformation of one kind of atom into another. The first such reaction squeezes hydrogen atoms together to form helium atoms, which is what our sun is doing all the time. This process creates heat and light, some of which escapes from the stars in spreading waves of photons. The burning gases on the outside of stars pull away in waves, like the skin a snake sheds, because of the gravitational pull of matter, such as other stars, around them. Stars must constantly keep their balance between tremendous forces pulling them apart and other forces squeezing them together.

Eventually, the first-generation stars collapse from growing so heavy they can no longer keep their balance between the internal and external pulling. Their atoms mass ever more tightly together. If the pressure inside the star gets too high, the star explodes, scattering stardust like seeds back into the galactic gas cloud. The mother cloud becomes ever thicker with the gas and dust of such explosions and gives birth to a new generation of stars as the old ones die.

The next generation of stars forges its atoms into yet bigger and heavier kinds until all the different kinds of atoms—all the different elements—of the universe have been formed from the original hydrogen atoms. Meanwhile, the central ring of gas clouds in a galaxy grows larger and more complicated, becoming a kind of skeleton that holds the galaxy together. At last many of the atoms from exploding stars are too heavy to form new stars and begin to form themselves into planets circling around stars that are made of the lightest elements. This is why our sun, although it is not a first-generation star, is made like one. The heavier elements of its parent star are in its planets.

So protogalaxies evolve into galaxies—whirling, weaving, squeezing, exploding, pulsing their insides into ever richer patterns and parts. Molecules formed of groups of atoms, even the kinds of molecules from which the familiar living systems of the earth formed themselves, are created in complex galactic processes, as we shall see later. For now, let us remember that the stars we see in our night skies are only a few of those in our own galaxy, and—as we see them with our eyes—they don't begin to hint of our galaxy's complex patterns and processes. Far beyond those stars lie billions of other galaxies, each made of billions of stars and planets wheeling in their clouds of gas and dust, creating who knows how much life.

○ ○ ○

Astronomers, whose name comes from the Greek *astron*—"star"—now know the different shapes of individual galaxies and can see them clustered into larger patterns. There are even clusters of clusters, called superclusters, even some greater pattern that extends all through the universe, parts of it appearing in our first images like huge curved strings and the holes in Swiss cheese. In our minds, these crude images, we may hope, will one day resolve themselves into an understanding of the greatest patterns of all.

Though we don't know what these patterns are as yet, it appears increasingly obvious that they form a cosmic unity of process and pattern rather than a chaotic spray of unrelated parts. A single notion that would account for such pattern is the concept of *mutual consistency*, which is at the heart of "bootstrap philosophy," a mathematical physics conception popularized by Fritjof Capra. This is the concept that the universe is a dynamic web of events in which no part or event is fundamental to the others since each follows from all the others, the relations among them determining the entire cosmic pattern or web of events. In this conception, all possible patterns of cosmic matter-energy will form, but only those working out their consistency with surrounding patterns will last. "Mutual" means shared; "consistency" is agreement or harmony. Thus we can sense mutual consistency as the shared harmony worked out among all cosmic patterns. The notion can be made more familiar by considering the shared social harmony worked out by groups of people when each individual adjusts his or her behavior to that of the others in a harmonious way. Anyone who cannot do this will tend to be excluded from the group, unless the deviant can force the others to make their behavior consistent with his or hers, in which case a new (if tenuous) mutual consistency would have been worked out. At present our species is not behaving in a way that is mutually consistent with the rest of our planet, and the consequences may not permit our survival.

Increasingly, then, we are discovering with our modern instruments and measurements what ancient peoples told in myth—that all of the universe is one great pattern, a single dance evolving into ever richer complexity over billions of years.

Until recently, scientists had a rather different idea of how nature forms itself—a mechanical idea of wholes built from parts, as machinery is built, though coming together automatically without any designer or builder. We shall learn more of this way of looking at things later, when we look at human history. For now what matters is to understand this new way of seeing that all evolution—of the great cosmos and of our own planet within it—is an endless dance of wholes that separate themselves into parts and parts that join into mutually consistent new wholes.

Between five and four and a half billion years ago, some of the gas and dust from that great star explosion gathered into an earth-ball made of twelve different kinds of atoms, or twelve elements. As it condensed, it grew heavier and spun around faster. The heat of pressure and nuclear reactions inside it melted the packed matter into a fiercely burning liquid. But the outside of this fiery ball, touching cold space, cooled off as a thin crusty skin, a bit the way homemade pudding forms a skin as it cools, or the way fat hardens on top of cooling gravy. The earth's skin was made of rock—a crust of rock around a hot, molten core.

While it was still very thin, this crust melted again and again, each time letting the heaviest metal elements sink back into the core while lighter elements formed a foam of rock around those fiery insides. Today's earth has a thicker crust, broken up into great *tectonic* plates that ride on the denser mantle surrounding a solid core. We can still see the hot liquefied elements of the mantle pouring out through volcanoes puncturing the crust. And in earthquakes we can feel the motion of the great tectonic plates as they slide about creating new geological formations.

In the myth of Gaia's dance, as her body forms mountains and valleys, the seas are formed from her warm moisture. Just so, it seems, the seas eventually pooled on the young earth.

At first, when the earth's crust cracked here and there, the liquid insides oozed out as lava. Lava, as the pressure that keeps it together is released, separates into heavy atoms that cool into more crusty rock, into water that hisses up as steam, and into other atoms light enough to float over or off the surface of the planet as gases. We now believe the water steaming off the hot crust stayed high above an early atmosphere of poisonous (from our point of view) gases for what may have been a long time, but eventually formed clouds that condensed into rain. The rain poured down so hard and for so long that the seas began pooling on top of the heavier rock. As more and more lava oozed through cracks in the earth's crust, the crust itself grew thicker and lumpier; as new clouds gathered and fell in cycles, the seas grew ever bigger and deeper.

As the earth's crust grew thicker, new streams of lava broke through it with greater force. Spitting volcanoes shot their fiery insides high into the air, forming mountains as the lava cooled and hot ashes settled down. More mountains were formed when earthquakes cracked the crust and slid parts of it over one another, and when the crust heaved and bulged without breaking. Rocks sliding

Cosmic Beginnings

We have already seen how the swirling gas clouds that evolved into galactic clusters began forming as soon as [...] joined together to form the first simple hydrog[...] particles universe thus evolved by forming m[...]en atoms. The early itself, many of them becoming ne[...]re and more parts within they proved consistent with other wholes surrounding them. As stars form within a protogalaxy, it becomes a galaxy—a great star system that in turn forms within itself relatively independent single or double star systems, some with planets such as our solar system. Later we will see how a planet's crust can form the packets of life we call microbes, or bacteria, and how these in turn can join together in building larger living cells, which in their turn evolve into larger creatures.

The universe of all these parts within parts, or wholes within wholes, reminds us of nesting boxes or of the Chinese or Russian dolls of various sizes that fit inside one another. The philosopher-scientist Arthur Koestler has suggested we call each whole thing within nature a "holon"—a whole made of its own parts, yet itself part of a larger whole. A universe of such holons within holons is, then, a holarchy—in Greek, a source of wholes—one original whole that formed ever more complicated smaller wholes within itself, some becoming holarchies themselves. We will use this image and the words "holon" and "holarchy" throughout this book.

○ ○ ○

Our own solar system, with its sun-star nucleus surrounded by planets, moons, asteroids, comets, and space dust, is a holon within the larger holon of our galaxy. It was born of the scattered gases and stardust of an older star that became a supernova exploding about five billion years ago. The earth is still so radioactive from this explosion that its core is kept hot by continuing nuclear reactions, and many atoms all over its surface—in rocks and trees and even in our own bodies—are still exploding. In our own bodies it has been estimated that three million potassium atoms explode every minute.

These explosions are much too tiny for us to see, feel, or perceive in any other way. They are not arranged to blow up neighboring atoms as well as themselves, as in our powerful man-made nuclear reactions. Still, they are evidence that stardust is not just fairy-tale magic; it is what we are really made of—we and everything else that is part of our world.

over one another were ground into sand and dust, like the dust on the moon and on Mars.

Huge dust clouds were created when meteors of all sizes—some of them as large as small planets—struck the earth, smashing into the crust, pitting it, breaking it up, mixing it with the broken space rocks themselves.

The gases floating around the planet, those just heavy enough to be held by its gravity, were nothing like the air we breathe now. There was no oxygen, but only a mixture of gases which, had the earth not come alive, would have eventually settled into something like the atmosphere on Venus and Mars today—an atmosphere without oxygen around a lifeless planet.

What, then, did the earth have that Venus and Mars did not? One of its special features has been called the "Goldilocks effect"—Venus was too hot, Mars was too cold, but the earth was just the right temperature for life. Another was its water, enough of it in liquid form in this "just right" temperature to carry supplies from place to place as blood is carried through a body. The constant transport of supplies must be possible for life to evolve.

Everything on earth—oceans and rivers, mountains and fertile fields, forests and flowers, creatures that float or fly or crawl or climb, everything, including ourselves—is actually made from the same original but recycled supplies, except for the small input of new meteors. Our world has created itself as new arrangements of the same atoms that started out inside a star, then formed the molten metal, crusty rock, and gases of a newborn planet—a planet that covered itself in seas as we have seen and is now ready to go on with its dance of life. Let's follow this great Gaian recycling system to see just how stardust continues to transform itself into a living planet—into all the amazing complexity of our beautiful world.

3

The Young Earth

Shall we think of the young earth at this point as a lifeless planet on which life is about to evolve? Most of us have been taught in school that animate matter is one thing—it is alive—and that inanimate matter is quite another, for it is not alive. Are the earth's rocky crust and watery seas inanimate, lifeless, matter while the plants and animals we know this story is leading up to are made of animate, or living, matter? Just what do we mean by the word "life"?

It may surprise you to learn that scientists do not agree on what life is. Some change their minds from time to time; others don't worry about the question "What is life?" believing the answer is known in some other science. In ancient times, when philosophers believed that all nature was alive, a physicist was someone who studied nature—*physis*—and so was concerned with living things. Later, when scientists decided to divide the world into animate and inanimate matter, physicists took on the job of describing how inanimate matter is put together, and biologists, whose name comes from *bios*—way of life—took on the job of describing living things.

Physicists think biologists know what life is because it is their job to know, but biologists keep changing their definition of life, and they pass the question of how to tell life from non-life on to chemists, whose name comes from ancient roots having to do with the transformation of matter from one kind into another. So chemists divide chemistry up, in their turn, into two kinds—

organic chemistry, the study of living matter; and inorganic chemistry, the study of nonliving matter. Chemists know something about the transformation of inorganic matter into organic matter, but the question of just when and where life began on our planet still gets tossed back and forth among them, or taken back to ideas from physics.

Some scientists nowadays thus talk about life in terms of *nonequilibrium thermodynamics*, to contrast it with the *equilibrium dynamics* of nonliving things—the physicists' way of solving the problem they created long ago when they declared that life was separate from non-life. Physics may not be the most appropriate branch of science to define life, but this new view at least talks about life as a process rather than as a kind of matter, and that seems closer to what life is all about.

Before religion and science parted company, the answer to the question of how life began was easy. Scientists themselves believed that God created living things—plants and animals and people—and put them into (or onto) the nonliving world he had created for them. But later, when scientists tried to explain the world without bringing God into the picture, they were stuck with believing that life is a special kind of matter that somehow comes from lifeless matter. One version of this belief was known as *spontaneous generation*—the belief that worms, for example, sprang from bits of dead garbage or rotting meat.

Louis Pasteur put an end to that, as we are also taught in school. Or did he? His very careful experiments showed that worms come only from worm eggs, and never directly from garbage. But where did worm eggs—which are living things—come from? Again the explanation seemed easy once we had the theory of evolution. They came from other worms, which had evolved from the smaller, simpler creatures we traced all the way back to microbes—living things so small they can be seen only through microscopes.

But where do microbes come from? That is still difficult to tell, but we assume they can be traced back to simple molecular systems that could reproduce themselves. Nowadays scientists believe that life began with small clumps or sacs of organic molecules. The organic molecules themselves were nonliving matter that came alive when they got stuck together in certain ways. In other words, life still seems to come from lifeless matter. In this sense, spontaneous generation was not so much disproved as pushed down to things much smaller than dead meat and worms.

We are still stuck with the question of just what life is. What is it that brings the lifeless molecules in some places, on some planets, to life when they are chained and clumped together in certain ways? Even though we are talking about very tiny things, there is still a big jump from nonliving matter to life.

We have already suggested that it might be better to see life as a process than as a kind of matter. Perhaps it would also help if scientists did not keep looking for the answer only in tinier and tinier parts of nature, believing that in doing so they would see just how things are built from the bottom up.

If we begin, instead, by thinking of wholes, or holons, that form their own parts from the top down, so to speak, everything looks very different. Think, for example, of the huge protogalactic cloud holons we talked about in Chapter 2. If we could watch a movie of the evolution of a protogalaxy sped up so that billions of years happened in a few minutes, what would we see? We would see it whirl and throb, grow and change, its parts dissolving and exploding, more complicated new parts forming in their place and even reproducing themselves as the mature galaxy took on its complicated form. Galaxies themselves split apart and merge with others on "collision." And within galaxies—perhaps within all of them—some planets produce what we all agree, here on earth, to be life.

While astronomers may speak of the lives of stars, they do not seriously count stars or galaxies as living beings. Yet galaxies do some of the things by which we all recognize living beings in our ordinary experience—such as keeping their form through many changes within them, creating and replacing their own parts, sometimes even growing and/or dividing to form offspring galaxies.

The most promising new definition of life among biologists, in fact, seems very nearly to fit galaxies, if not stars. This is the definition of life we owe to the Chilean biologists Humberto Maturana and Francisco Varela. Their concept of life is a process called *autopoiesis* (pronounced *auto-po-EE-sis*), which in Greek means "self-production." An autopoietic unity, or holon, produces the very parts of which it is made and keeps them in working order by constant renewal. An autopoietic holon works by its own rules and creates a boundary that distinguishes it from its environment and through which it exchanges materials with its environment. We do not see such boundaries around galaxies, yet galaxies are visible as distinct entities that maintain their shape while producing and reproducing their parts. The earth, as we will

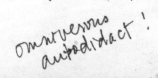

see, also produces and renews its parts, including the thick atmospheric boundary through which it exchanges radiation energy with its environment.

It seems that as we learn more about our universe, we need to change our scope and the questions we ask about life. Until now we have assumed that all the universe is nonliving matter except for some matter on planets such as ours. But why *should* we divide the universe up in this way? Physicists have now discovered, as we will discuss further in the last chapter, that the matter-energy of the universe right after the Big Bang—when we first know anything about it—was already, by its very nature, bound to form living systems. Had things been just the tiniest bit different at the beginning, this would not be so and we could not have evolved. Perhaps, then, life evolves as the essential process of the cosmos as a whole and is not just something happening at a special point we hunt for in vain. Perhaps galaxies are a very significant part of a cosmic life process. It certainly seems that our earth, born from it, is alive in its own right.

We do not know whether in our own solar system planets such as Mars and Venus began coming to life and then failed to evolve because they could not keep themselves alive. It is ever clearer that, as with the seeds and eggs of plants and animals, far more planets are produced than actually come to life. Planets must have just the right composition and be in just the right relationship to their star to come as alive as has our earth. Yet even if only a few planets among many succeed in coming to life, there must be billions of living planets in the universe. And the others, the majority of planets that do not come alive in their own right, may still play a supporting role in the life of their galaxies.

The creatures we are used to thinking of as alive, such as plants and animals, contain much supporting "nonliving" matter in their woody trunks and shells and bones, their thorns and hooves and nails, hair and scales. "Nonliving" planets may also be very much a part of live galaxies, perhaps even playing important structural roles in them. What about the earth itself? Many scientists argue that it cannot be a living being because only its outermost layer—thin as the dewy mist on an apple at dawn—shows signs of life. What, then, we may ask, about a redwood tree, which is 99 percent deadwood with just a thin skin of life on its surface? No one argues that redwoods are not alive.

It is new in modern science to look at the cosmos and the nature of our planet in this way. It is not easy for scientists to jump from seeing the earth as a nonliving planet that became a home for living creatures, to seeing it as a single living being with its creatures as much a part of it as cells are a part of our bodies. The scientific studies of earth have been divided, as we said, into studies of living and nonliving matter. Geologists have had the task of explaining how the geological "mechanisms" of nonliving matter, such as rock, change with time and weathering. Their work was not intended to be mixed up with that of the biologists who study living things, since these living things have been and still are believed by most scientists to arise in ready-made geological environments and either adapt to them or die out.

Now, however, the jobs of geologists and biologists are getting mixed up whether they like it or not, for the same stardust that was transformed into a rocky planet continued to be transformed into living creatures. What we are made of was stardust at one point, that is, but we can also see ourselves as made of the rock which that stardust became, as part of the earth!

To make things more complicated, much of the rock that is transformed into live creatures is later transformed back into rock. And so, just as creatures are made of atoms that were once part of rock, almost all rocks on the earth's surface are made of atoms that were once part of creatures—creatures that built themselves from the atoms of still earlier rocks.

Think about that. The recycling of stardust gets to be a complicated matter as a planet comes to life. Geologists are now just beginning to believe the Russian scientist V. I. Vernadsky, about whom we will say more later, who understood life on earth as "a disperse of rock"—rock rearranging itself over billions of years, rearranging itself into ever more complicated forms of life from microbes to men.

That alone is enough to mix up geology and biology, but there is even more to it. Our planet never was a ready-made home, or habitat, in which living creatures developed and to which they adapted themselves. For not only does rock rearrange itself into living creatures and back, but living creatures also rearrange more rock into habitats—into places comfortable enough for them to live in and multiply.

But let's take it one step at a time and look first at life as rock rearranging itself. How can this happen?

We begin to see that there is more than one way to understand what life is. We just saw it as a mixture of geology and biology. Let's now try looking at it as a mixture of physics and chemistry.

Remember the forces, such as gravitation, that helped create patterns in the cosmic dance of particles and atoms? One of those forces is the electric force that holds atoms together. This force keeps the outer particle dancers of atoms—called electrons—from flying off into space away from their nucleus of heavier particles. This is not entirely unlike the way gravitation keeps planets from flying off into space away from the sun, though the orbiting electrons are not hard balls like the ones on the old-fashioned atom models that looked like the solar system.

At the macrocosmic level, powerful electrical, or magnetic, fields were set up by the interaction through the earth's crust of the sun's energy and the molten metal in the earth's core. We might compare this with a giant battery whose energy can be used to do all sorts of work. At the microcosmic level, the electric force allows electrons to dance in two atoms at once, holding them together as a molecule. The more atoms that dance together in this way, the larger the molecules formed.

The strong energy of sunlight coming to the earth's crust through the thin early atmosphere stirred up the molecular electric force within the great electric fields, creating storms above and breaking up molecules in rock dust, mud, and seawater below, re-forming them into new and larger molecules. When molecules break up and recombine in new patterns, we call it a chemical reaction, since chemistry is the study of such transformations in the patterns of molecules. The energy that stirred up the electrical force recombined many molecules of the earth's crust.

Such chemical reactions also happen elsewhere in our galaxy. The larger organic molecules such as those of sugars, acids, and lipids (fats) that were formed on the young earth are also formed in large quantities and great variety somewhere in the center of our galaxy and perhaps all over it. Some of them come to earth by way of meteors. It is even possible that those "planet eggs" which come to life may be "fertilized" by meteors.

Some chemical transformations, as we said, were due to electrical storms created among the clouds of cooled steam in the early atmosphere as the sun's energy heated the earth's surface. Besides helping large molecules to form, these storms drove a water recycling system, collapsing clouds into rain, which fell on land and sea, the water rising again by evaporation and collecting back into clouds.

Rainwater ran over the rocks, creating grooves that formed riverbeds and valleys, carrying ground sand and dust full of rock salts to the seas. Rivers and streams thus formed as the bloodstream of our embryo planet, carrying the supplies needed to develop or evolve its life. For a live planet needs not only a great deal of energy but also flowing matter such as atmospheric gases and water to move things about. As we will see, planetary life is not something that happens here and there on a planet—it happens to the planet as a whole.

The largest new molecules probably formed in shallow waters with the help of sunlight and lightning storms—or perhaps with the help of the earth's internal energy around cracks in the earth's crust on the sea floor. Even the sun's drying heat at the water's edge may have played a role in forming large molecules and packaging them. Large molecules absorbed a lot of electrical energy, which was then useful in speeding up their chemical reactions to form ever larger molecules—giant molecules built from the large molecules of sugars and acids. Some scientists believe the giant molecules formed as large molecules lined themselves up on molds or templates of clay or other crystal matter that had regular, repeating surface patterns or notches for the molecules to hold on to. Others believe that the production of giant molecules happened only after the earliest molecular life systems were already organized within tiny capsules.

The central chemical stuff of Earthlife—that is, of the autopoietic holons forming within the earth holon—is carbon, or rather reduced carbon compounds, which are carbon atoms surrounded by hydrogen atoms. The lively energized carbon of the earth also combined easily with oxygen, nitrogen, sulfur, and phosphorus to form all sorts of organic molecules and substances. In fact, you are made of very little other than these six elements in their rich variety of combinations.

Among the giant molecules formed from smaller ones were *proteins*—long strings of *amino acids,* which are themselves molecules made of various combinations of a dozen or fewer carbon, nitrogen, hydrogen, and oxygen atoms. Other giant molecules, assembling from both acids and sugars, were those we call *ribonucleic acid* (RNA) and *deoxyribonucleic acid* (DNA). DNA may actually have been a later development of early living systems based on RNA. Whatever the exact sequence, DNA and RNA came to work together with proteins as the copying and building system of life.

DNA molecules are long chains of smaller molecules joined into long, twisted zippers. We have discovered that the "teeth" of these DNA-molecule zippers act as a four-letter code that can be arranged in endlessly different patterns, just as letters of the alphabet can be written into words and sentences and books. Because of this, a DNA molecule can hold information. After all, information is anything "in formation"—anything that is in an ordered pattern rather than in chaos. (Some scientists argue that whatever is in formation is information only when used by a living system; in this book we define information as anything that is in formation—a book, for example, contains information even if no one reads it.)

Information, if it can be copied, can be a plan—a plan for a new copy of itself, or a code plan for something else. Now, DNA can copy, or replicate, itself, but not without the help of proteins that can unlock DNA zippers. Once unlocked, the DNA unzips itself into two half-zippers. As these float around in a soup of smaller molecules, the "teeth" of each half—all letters of the DNA code—attract new partners just like those that were opposite them in the closed zipper, because those are the only ones that fit into place.

Presto! We have two zippers where there was one—and the two are exactly alike if no mistakes have been made. The DNA-protein partnership evolved in such a way that while proteins unlocked DNA zippers, they also "persuaded" DNA to store plans coded for building more protein as well as more of itself. Because DNA could be replicated, the DNA-protein partnerships were capable of reproducing themselves.

This is a bit oversimplified, since viruses, which contain only RNA or DNA coated by protein, have to get inside cells where other things are available in order to reproduce. Nevertheless, protein with DNA or RNA, or both DNA and RNA, formed molecular cooperatives that became the basic reproduction system of carbon-based life. This *genetic* system is usually described as one-way, the DNA code strictly determining the production of proteins, which are the main building materials of living holons within the earth holarchy. But recent evidence indicates that proteins can in turn affect and change the DNA code. We will get back to this form of cooperation in later chapters.

At some point early in the earth's history there were plenty of the sugar and acid molecules that were needed to build the long chain molecules of RNA, DNA, and protein. And so the formation of cooperative partnerships very likely became inevitable in the earth's warm wet mud and shallow seawater where molecules could float about and bump into one another. Possibly there was a long time when these partnerships could hardly have been told apart from the thick soup of building materials around them.

Some scientists, however, argue that such partnerships really could not have gotten under way until the molecules were enclosed in sacs, or membranes, that held them together with other supply molecules and protected them from being dissolved. The most likely candidates for such sacs are called *liposomes*—meaning "fat bodies." Liposomes—so tiny they can be seen only with an electron microscope—form as hollow spheres of lipid molecules, something like microscopic soap bubbles, whenever lipid molecules find themselves in water. This is because the tails of these lipid molecules are *hydrophobic*—swinging quickly away from water, protecting one another from it by turning inward so that their heads form a tight sphere around them. Sometimes a double-layered sphere forms with water inside and outside, the double layer having all the lipid molecule heads on both surfaces, with all tails between the two layers of heads. This is the typical formation of simple cell walls and persists even in the most complex cells today.

If a soup containing liposomes and a variety of large molecules is repeatedly dried out and liquefied again, the liposomes break open and flatten out during dry times and re-form their spheres in wet times, often around large molecules—even as large as DNA and protein molecules—that have become trapped inside them while they were broken open. Such conditions must often have occurred at the edges of early seas. The liposomes themselves then function as a skin, or membrane, which serves the molecules inside it both as a protection from and as a connection to the outside world. This membrane permits selective chemical crossings, allowing some kinds of atoms or molecules to come in and other kinds to pass outward through them. This soon makes the inside environment chemically different from that outside. Such an arrangement fosters the development of chemical cycles that are basic to living cells.

However the first cells formed, protein became the main material of which living creatures built themselves, while RNA and DNA stored the plans and made it possible for living things to multiply. Some protein molecules came to play a particularly important role by speeding up what other molecules did—say, speeding up the chemical reactions that build new protein or copied DNA. We call these special proteins enzymes, and their wonderful talent for speeding up the chemical dance is very important to our planet's life. In fact, the presence of enzymes has been suggested as one way of defining the presence of life, and the first enzymes likely occurred as a widespread chemical earth event, perhaps both outside and inside early cells.

While details are still missing, this is essentially how the solid and molten crust of the earth began to rearrange itself into living creatures. Some of its material gassed off into atmosphere, part reformed into seas, some broke up and was washed into the seas. With the help of great amounts of energy, larger molecules formed and joined into partnerships, set up chemical cycles in early liposomes, speeded up their own reactions with enzyme activity, reproduced themselves, and through all this established themselves as living, or autopoietic, holons—early creatures in their own right. These creatures dwelt within the larger living holon that had given them life and to which they gave a new kind of life in turn. Thus on the one hand we can say that tiny separate living holons evolved all over the earth, but on the other hand we can say that the earth holon was coming ever more alive as it evolved its own autopoiesis through a new kind of self-packaging chemical activity.

○ ○ ○

From our old point of view we could see the beginnings of life only as a collection of microbes descending from some primeval cell that formed accidentally somewhere on earth, giving rise to offspring that were forced to adjust or adapt their way of life to it by natural selection, which we will discuss in Chapter 7. This was a logical way to see things when we formed our concept of life from our study of individual creatures small enough for us to see as wholes. In our new way of seeing life as autopoietic systems that may be as large as the earth or even larger, we can think of Earthlife as a planetary process—as the chemical reactions of the planet's crust speeding up, transforming the crust into masses of microbes, which in turn transform more of the crust into a livable home, as we will see in the next chapter. And while all this happens at the microcosmic level, the macrocosmic events of the largely molten, still radioactive planet keep its crust heaving, cracking, and sliding, pushing up mountains, buckling in valleys, changing the shapes and positions of continents amid its deepening seas. All together, this is the self-creating dance of a living planet driven by its sun and by its own energy.

One way of looking at all this is to see the earth as having come alive through all sorts of "border activity." The crust that stirred to life was the boundary enclosing the earth and at the same time connecting it to outside energy from the sun and to new materials coming in as meteors. The first cells seem to have formed specifically at the boundaries separating and connecting the land and the sea. These cells' own boundaries made their individual lives possible by separating them from and connecting them to their environment. At all levels from great to small, this border activity can be seen as highly creative—a lesson we humans, with the boundaries we have created among ourselves, might well take to heart.

Let's now imagine that we are watching a fast-running movie of the early earth as it evolves within the larger being of our Milky Way galaxy. As we approach the earth, we see it whirling and heaving, its thin crust rising and falling, breaking and slipping, bleeding lava where it tears open and sighing bursts of steam. Meteors and planetoids, which are still part of the supernova's debris, strike and wound the earth, making great splashes of molten rock and gas; the thin atmosphere is often reddish with smog produced by the reactions of its own gases. Lightning flashes, and seas form during heavy rains until masses of land and sea become distinct, though the seas are brownish beneath the murky atmosphere. Slowly the crust thickens and cracks into plates that slide slowly over the surface, carrying the land masses into new patterns. Patches of colored microbes appear and grow along the shores; gradually a tougher but clearer atmospheric skin develops, making the seas turn a sparkling blue. Meteor impact is low; turmoil subsides, and much of the land becomes covered in green. Now and then ice moves down over the green before withdrawing again to the poles, raising and lowering the level of the seas, covering and uncovering the land as though the whole planet is breathing in some gargantuan rhythm. Everything is in constant motion as the earth shimmers and glows in the sun against the darkness of space, its changing cloud patterns swirling over blue seas and varicolored lands.

These changes actually happened over billions of years, at a rate too slow for us to recognize as very active. Yet a billion years is less to our planet than a decade is to us. And when we use our imagination to see these changes within the time span of a short film, our planet looks very much like a living creature. Our movie makes the young planet appear to be trying hard to express itself in a new way as its materials churn about, its crust forms and reforms, its seas and clouds pool over the rocky crust. It has enormous energy of its own and receives more energy from the sun, which sends it light and heat. It might remind you of X-ray films of the transformation of a caterpillar into a butterfly inside a cocoon, or a chick embryo turning and growing inside its shell.

Already at this early stage the earth begins to fit the autopoietic definition of life as it is creating its own parts, including the tiny autopoietic microbes which, as we will see in the next chapter, create the thickening atmosphere that becomes a new boundary membrane or skin. In later chapters we shall see more evidence of autopoiesis as new complex holons form within the planet's holarchy.

Had our movie shown the other planets as well, we would have seen the sharp contrast as they settled into drabber orange-brown inactive orbs while Earth's tremendous, energetic activity brought it to life with radiant blue-green colors and its swirling breath of white cloud.

4

Gaian Problems

Imagining Gaia as a beautiful goddess dancing gives us a poetic metaphor for nature's living beauty. But real life is often hard and troublesome, as we know from our own experience. And Earth had big problems right from the time its Gaian dance of life began. In more scientific terms, we might say the probability that Gaia—our name for Earthlife as a whole—would continue to evolve was rather low during its early stages, or that a stable autopoietic Gaian system evolved only under considerable threat to its existence.

At times when telling the story it is easier to speak metaphorically about Gaia—or about her bacteria, for that matter—in terms of our own experience with problem-solving and invention. The reader should not find it difficult, however, to distinguish between such metaphoric shorthand and the underlying scientific account, which includes no assumption that Gaia—or bacteria—think or plan ahead, or feel their failures and successes as emotions.

Even when its crust was already coming alive with microbes, the young earth, whirling more than twice as fast as it does now, still hissed with steam, cracked so that its lava flowed like blood, and was endlessly bombarded by meteors belting in through the thin atmosphere, raising dark clouds of dust as they wounded its still tender body. The embryonic earth's continuing life was not at all a sure thing; Gaia was not yet a secure, stable being able to maintain itself.

GAIAN PROBLEMS

The constant hail of meteors, leaving craters such as we see on the moon, was a serious threat. Though meteors may have contributed important molecules such as lipids to the formation of microbes, they might also have killed them off again. Every day these space rocks of all sizes came hurling from the sky like bullets. If nothing had happened to protect the earth from them, it might well have ended up as lifeless and pockmarked as the moon and our neighboring planets.

There may also have been another problem, though scientists differ on this matter. The sun's energy was most helpful in splitting molecules so that new ones could form, but as the first microbes formed and multiplied, the strong sunlight may have been too much for many of them to stand, putting them in need of protection from the burning part of sunlight we call ultraviolet radiation. Some ultraviolet is good for living creatures, but too much can burn them, and our young sun probably produced far more ultraviolet than it does today, when we are concerned about our own threat to the shield of ozone protecting us from it—a shield that did not exist at all around the early earth, though the smoggy early atmosphere may have offered some protection.

In any case, the first microbes seem to have formed in seawater or wet mud just deep enough to filter out the dangerous rays. There, as we saw in the last chapter, bits of a rich soup of organic molecules and seawater were probably trapped in liposome spheres where the molecules could move about and begin new kinds of chemical cycles. These would have included—had they not already been formed elsewhere—the construction of the giant RNA and DNA molecules that became useful as a storage system for life's plans. In the self-production and reproduction cycles that gradually evolved, RNA lined up with DNA to copy its information, then lined up with amino acids to produce the proteins coded for, which in turn helped DNA split apart to copy itself, and so on around the loop. But giant RNA and DNA molecules could be broken by ultraviolet light, and so one of life's earliest inventions, not long after reproduction itself, was the repair of DNA with special enzymes.

Lipid walls enclosing the early microbes—now becoming full-fledged bacteria—permitted the entry of new raw materials and the disposal of wastes. Every living being or system has to cycle and recycle supplies. As Gaian weather cycles circulate water, for example, from sky to ground and sea and back to sky, rock is dissolved in running water and swept to the sea, gases are cycled, and the planet's temperature is regulated to a large degree through a constant balancing of gases with the amount of cloud cover. Meanwhile, as we will see in more detail later, dissolved rock used in forming the bodies of sea creatures ends up buried on the sea bottom, which later spits the new rock formed by them back up as land to begin the cycle again. Just so, the liposome microbes formed in the earth's crust developed internal cycles for circulating their own supplies and carrying out the business of life.

These ancient evolving microbes gradually replaced their tiny spherical capsules with larger, more flexible cell walls and evolved into bacteria. They still depended on seawater to float supplies to them, or to float them to supplies, and to float away wastes they could no longer use. By trial and error they learned to use these supplies to grow themselves, to repair themselves when they suffered damage, and to reorganize themselves as needed, keeping records of their new discoveries in their DNA.

Every living creature must get materials and energy from its environment to form itself and to keep itself alive. What is left of these supplies after the useful parts and the energy have been taken from them, along with whatever else was part of the creature but is no longer of use to it, is waste that must be gotten rid of—by returning it to the environment. This is why no living creature can ever be entirely independent—it is a holon within a larger holon, depending on its environment, which *is* its larger holon, for its very life.

As Koestler put it, a holon has at once the *autonomy*—in Greek, "self-rule"—of a whole and the dependence of a part, since it is embedded within larger holons on which its existence depends. Koestler grappled with this concept of dependence-in-independence, referring to it as an "integrative tendency," or even as "self-transcendence." The most appropriate term is *holonomy*—the rule of the greater whole or holon that must be balanced with self-rule, or autonomy. The physicist David Bohm uses the word "holonomy" in exactly the same sense when describing how the autonomy of subatomic particles is stabilized by the rule of all other particles around it—by holonomy.

Any holon containing smaller holons—such as an earth full of bacteria or a body made of cells—tempers the individual autonomy of its components with its *own* autonomy, which is *their holonomy*. Any individual human, for example, must transcend simple self-rule and integrate him- or herself with the rules of society, while human society must transcend its autonomy and integrate itself with the holonomy imposed by the autonomy of the planet. The balance between any holon's autonomy and holonomy must be worked out as mutual consistency if the holon is to survive as part of a holarchy—and it cannot survive in any other way if we accept the fundamental notion of mutual consistency as described in Chapter 2 and as illustrated in later chapters.

Bacteria are holons within a larger holon consisting of all bacteria (more on this later) and within their environmental systems, or *ecosystems*. While we are talking definitions, let us call systems of organisms "ecosystems," to distinguish them from "biosystems," which may be individual *or* collective organisms. Bacteria are also called monera—the first kingdom of living things in our present classification. Each moneron is just one walled cell, and yet it is also a whole organism or creature. The tiny monerons that were Gaia's first creatures were thus the first more or less independent holons within the earth holon.

Fortunately for these early monera, the sea was full of supply molecules, ranging from small dissolved rock salts to the larger sugar and acid molecules needed to build DNA and protein. So the bacteria could grow and divide and grow again, spreading themselves thickly throughout the seas. Even today, bacteria, or monera, are by far the most numerous creatures within the great Gaian organism.

As the bacteria multiplied, winds and water driven by the sun's energy swirled the rich chemical soup about, stirring it into ever greater activity. The more bacteria there were to suck up supplies and blow out their wastes, the more the whole chemistry of the earth changed—sometimes the worse for life, sometimes the better, as we will see.

The early monera were getting their energy by breaking up supply molecules in a process we call *fermentation*. The bacteria we use to make cheese, yogurt, and wine still work the same way today. Yeasts, such as those we use to make bread, do it, too. Fermenting bacteria can be thought of as "bubblers," since they make bubbles of waste gases, like the bubbles you see in risen bread and in cheese. Wherever you see bubbles rising in mud or natural waters, fermenting bacteria are probably at work.

Breaking up molecules—by fermentation or in other ways—frees the energy that held them together. The bubblers stored this energy in a special kind of molecule we call adenosine triphosphate (ATP). At first they may have found ready-made ATP molecules in their surroundings, but eventually they learned how to make them. The bubblers kept the energy-loaded ATP handy until the energy was needed for building, repair, and other work. Every

living thing on earth since then has been using the ATP energy storage system invented by the bubblers, though bacteria later discovered faster, better ways of making ATP than by fermentation.

In addition to energy, of course, the bubblers needed building supplies, and for a long time, as we said, large sugar and acid molecules were plentiful in the environment, ready to be split up or used as they were. To reproduce, some monera copied their DNA and then split themselves down the middle in the process we call *mitosis,* building two offspring monera from their own split halves. Others budded off smaller bits of themselves containing copied DNA to start their offspring. When supplies got low here and there, some bacteria learned to pack their DNA and a bit of protein into solid little spores with tough shells. These spores floated about doing nothing at all till they came to places where supplies were plentiful and they could grow into proper monera.

○ ○ ○

Over time, monera built new kinds of protein and new enzymes and invented new chemical processes and cycles, new parts for themselves, new lifestyles. More than three billion years ago, then, bubbler monera were multiplying and dividing into different strains, forming a thick soup or surface scum, living off ready-made supplies of large sugar and acid molecules. Some strains of bacteria learned to use the acid and alcohol wastes of others, and to set up efficient cycles of using one another's wastes as supplies. Some learned to make the nitrogen of the atmosphere usable by combining it with other elements. Had they not, life would have died out from nitrogen starvation, as nitrogen is one of the six basic elements needed to build living things.

Still, as competition for large-molecule food supplies increased, a new crisis developed. As if Gaia didn't have enough problems already, it began to look as though her first tiny creatures might die for lack of supplies.

But they didn't. Life is far too inventive to give up so easily.

What happened to the monera back then is rather like what is happening to us humans today. We have been making much of the energy we need to live in our human societies from the coal and oil supplies found ready-made in our environment. Now these supplies are running out, and we must find new ways to produce energy. A very important way—and probably the best way—of doing so involves the use of sunlight, or solar energy.

This is exactly what some monera began doing as their supplies ran low. Some elements they had to have in order to build their living bodies were all around them, but like the atmospheric nitrogen, they were not in usable form. Others were hard to get at, such as the nitrogen locked into the salty nitrates of the sea or the carbon locked up in the carbon dioxide gas of the atmosphere. There was plenty of carbon and nitrogen all around, but the bubblers had to invent special ways to unlock the carbon and nitrogen and then "fix" them—turn them into usable body-building molecules.

Perhaps the bubblers' most important discovery was finding ways to trap sunlight and turn it into ATP energy, which they did by using certain light-sensitive chemicals such as the porphyrins that make our blood red and the chlorophyll that makes grass and leaves green. They could then use this energy to split molecules of carbon dioxide gas, water, and rock salts into atoms, which could be rebuilt into food sugars, DNA parts, and more ATP for the work of growing, repairing, and reproducing. This process is, of course, *photosynthesis*—the "light-making" or use of light in the manufacture of food.

Some of the photosynthesizing monera are called *blue-green bacteria* because of the color their photosynthesizing chemicals gave them. Let's call them "blue-greens" for short. Their new way of life was very successful, so they multiplied quickly. After all, the blue-greens, unlike the bubblers, needed no special supplies. Water full of dissolved rock salts was what they lived in, and the atmosphere was full of light and carbon dioxide.

There was only one problem: their wonderful new way of making their own food and energy was also creating pollution.

Both the bubblers and the blue-greens made waste gases as they worked, but light-making food from water and carbon dioxide gas produced a very poisonous waste—so poisonous that it killed living things. This poisonous waste gas was *oxygen!*

We are used to thinking of oxygen as good and necessary, as a life-giving and life-saving gas that we breathe. But for the first living creatures, it was deadly. It is oxygen that turns metals to rust and makes fires burn. Oxygen destroys the giant molecules of living things, burning them up just as ultraviolet radiation and other kinds of radiation do. In fact, oxygen is more destructive than ultraviolet, for the large molecules needed to build the first living things could never have formed if the atmosphere had been rich in oxygen then, as it is now. So, when the blue-greens began making oxygen, they began making trouble.

Every molecule of carbon dioxide is made of one carbon atom and two oxygen atoms—"di" meaning "two." And every molecule of water is made of two atoms of hydrogen and one of oxygen. It takes six molecules of carbon dioxide and six molecules of water to make one molecule of food sugar. But when the sugar molecule is built—from carbon, hydrogen, and oxygen atoms—there are six molecules of oxygen gas left over as waste.

This is the oxygen that began polluting the early earth after photosynthesis began. At first the free oxygen combined harmlessly with dissolved rock minerals such as iron, making them rust, and built itself into rock; then it began piling up in the atmosphere.

It was as if a giant pump had been turned on in the seas. Bacteria were pumping carbon dioxide out of the atmosphere, using the carbon and pumping oxygen back into it. They also pumped nitrogen out of nitrate sea salts, "fixed" some of it for their use, and pumped useless nitrogen gas into the atmosphere. The living earth was bringing its own special atmosphere into being. Our nearest planet neighbors, Venus and Mars, have atmospheres made almost entirely of carbon dioxide, just as Earth's very likely was when this great pump got going. But now our atmosphere is almost all nitrogen and oxygen—because life made it so.

But how did life survive the poisonous oxygen?

Much of it didn't survive. Some kinds of bubblers that didn't need to be near light dug themselves down into mud where the poisonous oxygen could not get at them. Their fermenter descendants still live today by hiding from oxygen—in mud or in other safe places such as the stomachs of cows, where they help digest hay, or the beadlike root nodules of peas and beans, where they fix nitrogen to enrich the soil. But many, if not most, kinds of early bacteria must have been killed as oxygen piled up around them.

○ ○ ○

What was Gaia to do? Her dance of life had produced a rich array of living bacteria despite the dangers of meteors and ultraviolet light. Now most of the early kinds were dying just because some had discovered a new and better way to live. It was a lesson Gaia learned more than once, that new experimental forms of life may seriously endanger the whole dance and that other improvisations may be required to rebalance it. Not that Gaia could experiment on purpose or think about the results; nevertheless our evolving planet developed (and still develops) its "body wisdom" through experience, just as our own bodies within Gaia evolved the body wisdom by which they run and maintain themselves.

Though clever species of bubblers survived, they were no longer the main kind of monera. Blue-greens invented enzymes, which made the oxygen they produced harmless to themselves, and which took over. Some also learned to make ultraviolet screens—as we make sunglasses and chemicals to protect ourselves from sunburn. These were able to live successfully in stronger sunlight, where they could make plenty of food.

Others solved the problem of ultraviolet burn by living together in thick colonies. Those on top were burned to death, but the dead cells made good filters, absorbing the burning rays while letting the rest of the light reach those that needed it below. This was another way in which some lives were given for others, and a good reason for bacteria to live as cooperative life teams rather than as independent individuals.

You can see such colonies of bacteria beginning as a greenish brown scum on damp walls or muddy ground. Near the sea, they trap sand and other particles, forming thick muddy masses in shallow waters as live bacteria multiply and keep climbing toward the top. In some places we can still see this mass harden into rocks called *stromatolites*. In ancient stromatolites, the bacteria that have turned to rock can still be seen and identified.

The number of such rocks that were formed billions of years ago, and that are still being formed in the same way today, tells us just how successful the oxygen makers have been. Those rocks also show us how rocks that rearranged themselves into living creatures can rearrange themselves back into rock.

For about two billion years—almost half of Gaia-Earth's life—the blue-green oxygen makers were her most successful creatures. They multiplied into thousands of different kinds all through her water and mud, making more and more oxygen. And then they made yet another dramatic discovery: they learned how to use the waste oxygen they created along with food molecules to burn those food molecules for energy.

This process of burning food with oxygen is what we call *respiration*—the third way of making ATP, after the fermentation of the bubblers and the photosynthesis of the blue-greens—and the most efficient way of all! In respiration, the destructive energy of oxygen is used to break up food molecules and thereby free both their parts and their energy for use. It is a much more powerful way to do this than fermentation. Soon other kinds of bacteria learned this method of using poisonous oxygen to good advantage. Since we call the intake of oxygen for breaking up or burning food molecules "breathing," let's call the kinds of bacteria that use respiration "breathers."

○ ○ ○

Before we go on to discuss the changing world of the bubblers, blue-greens, and breathers in the Great Bacterial Age, let's look at how bacteria managed to accomplish so many innovations and lifestyles. Simple as bacteria are in comparison with later evolved creatures, they are remarkably ingenious, and we still have a great deal to learn from them. One of our own greatest new advances in science is something we learned from them only recently—*DNA recombination,* the secret of bacterial success. The bacteria discovered this process billions of years ago, and we are only just now catching on.

Until recently, scientists believed that changes in the forms of living creatures could happen only as a result of accidental changes in their DNA plans—by mistakes in copying DNA, for instance, or by accidental breakage and recombination of DNA at certain times, such as when it is struck by fast-flying nuclear particles that zoom through the atmosphere and through us without our notice. But now scientists are finding that some DNA changes look anything but accidental.

Modern bacteria have obviously been able to change very quickly in ways that protect them against the poisonous antibiotics we humans use against them. To do this, they have to make changes in their DNA plans. Life, it seems, does not just wait around for lucky accidents to solve problems and improve things, but is quite inventive, especially under the pressure of troubles.

We can easily see with modern microscopes that bacterial DNA is a very long molecule formed into a floating loop inside the tiny creature. We can also see that bacteria coming close to one another often dissolve parts of their cell walls and exchange bits of DNA. This means they are exchanging plans, or information—making new combinations in their DNA plans.

This information exchange, or communication system, of ancient (and modern) bacteria is at least as remarkable as any of their other inventions and probably made most of them possible. We are just beginning to learn how it works and to recognize it as the innovation we know as sexual reproduction, which we assumed until recently was discovered much later by much larger creatures.

Sex is the production of creatures by a combination of DNA from more than one source. Every time bacteria receive bits of DNA from others, they are engaging in sex—making themselves the product of two bacterial sources even though they are *not* reproducing. This sexual communication system apparently belongs to virtually all bacteria of all strains, so that bacteria can trade their DNA plans—or genes—with one another all over the earth!

In fact we begin to recognize that all bacteria form a great holon with a common pool of DNA genes—a single live network or system that covers our planet and through which the bacteria can trade and recombine genes according to need. Their world bank of DNA information may even be useful to larger creatures, as we also see bacteria coming into plants and animals to trade bits of DNA. Even before we made this discovery, we knew that no other form of life could survive today without bacteria. Why this is so will become clear as we watch the dance of life develop.

The bacterial gene pool of the young earth made it possible to spread resistance to oxygen through various protective devices, as well as to spread its use for breathing, or burning food molecules. Of all Gaia's creatures, the blue-green breathers were the most independent ever to evolve, and they are still going strong today, billions of years later. They make their own food and burn it, using only the simplest supplies. If they drift away from light, they work in the dark. They fix their own carbon and nitrogen. Living in water, they do not even risk drying up, as do the land plants that evolved long after them to carry on the double lifestyle that the blue-green breathers invented.

The bubblers, the blue-greens, and the breathers had now invented the only three ways that living beings all over the earth, even today, make their ATP, or their body energy. Like fermentation and photosynthesis, respiration produces waste gas. But this time the waste gas is carbon dioxide—the very gas needed for photosynthesis. What an incredible new opportunity. When food was made from water and carbon dioxide was produced by photosynthesis, a waste supply of oxygen was left over. This oxygen could now be used to burn the food for energy. Thus respiration completed a cycle by leaving a "waste" supply of carbon dioxide with which to start photosynthesis anew.

Again we are reminded of lessons people are learning today. First the ancient bacteria solved their energy crisis by learning to use solar energy; then they discovered that recycling supplies is a wonderful way not to run out of them.

The recycling of carbon dioxide and oxygen thus began within the simplest and tiniest Gaian creatures—creatures that multiplied with such great success that their recycling system has been an essential part of the Gaian life system ever since. In time, the two parts of this cycle—photosynthesis and respiration—became ways of life for different kinds of one-celled creatures, as we will see. Then, much later, plants and animals evolved to cooperate in producing carbon dioxide and oxygen for each other's use—or in recycling each other's waste, depending on how you look at it. But bacteria alone were, and are still, quite capable of managing the life of Gaia without help from more complex creatures.

As the oxygen piled up in and thickened the atmosphere, it not only created new problems requiring new solutions but itself became the solution to some old problems. Destructive as oxygen was to so many kinds of microscopic monera, it now helped in forming a protective blanket of air around the living earth. Just as in our ancient myth Gaia first formed the seas in her dance of life and then created a protective atmosphere, so it was in reality.

This blanket of air seems very thin to us. We can just barely feel it by waving our arms around in it. But what we feel against our arms would be much harder if our arms were waving much faster. Meteors move so fast that the air is quite solid to them. And rubbing hard against something solid produces heat, as you can easily demonstrate by rubbing your hand hard over a table. Meteors rub up against air so hard that the heat, together with the oxygen, ignites them and burns them up. The more oxygen there was, the more meteors burned up, until so few of them got through the atmosphere that life was much safer on earth.

Nor were meteors the only outside dangers oxygen protected Gaia against. The oxygen in our air is made of twin oxygen atoms dancing together as free-floating molecules. As ultraviolet rays strike these molecules, they break up the pair, leaving separated twins to join other oxygen pairs as triplet molecules. Such triplet molecules are no longer oxygen gas; they are *ozone*. And it is very difficult for ultraviolet rays to pass through ozone because it absorbs them. When there was plenty of oxygen, a whole layer of ozone collected in the middle of the atmosphere, shielding the earth from dangerous amounts of ultraviolet radiation.

Microbe-produced oxygen probably even played a role in preventing the seas from drying up, because atmospheric oxygen can trap evaporating lightweight hydrogen as water, thus preventing it from escaping into space and allowing it instead to fall back into the seas in the form of rain. Methane-producing fermenters may also help hold the oceans onto earth, as atmospheric methane decomposed by ultraviolet rays creates the *tropopause lid,* which is another barrier to the escape of hydrogen into space.

○ ○ ○

From this early history of the Gaian dance of life we can see that great problems are great challenges, and that living things are very inventive when faced with challenges. Maybe that is one of the most important things we can learn from Gaia.

It remains to be seen whether we humans will prove as creative as ancient bacteria in the face of the problems *we* make. Over billions of years, most of the carbon dioxide was pumped from the atmosphere, while nitrogen, oxygen, and rarer gases produced by living creatures filled it. Over this long time, life worked out exactly the right balance of gases that was best for it. Now we are changing that balance in dangerous ways.

Our use of coal and oil, for example, is creating a very serious double problem. Not only are we using up these important fuel supplies but we are also polluting the atmosphere with too much carbon dioxide in burning them. Coal and oil are made of ancient forests that built much carbon into their plant life. As they were pressed underground over time, the carbon was buried and transformed into natural fuels. When we dig them up and burn them, the carbon is released back into the atmosphere as carbon dioxide.

At the same time, we are also burning today's growing forests to clear land for our use. This releases even more carbon into the air while killing the very plant life that uses up carbon dioxide to make oxygen, thus preserving the balance.

Billions of years ago oxygen was the great danger. Now the danger is too much carbon dioxide. The result, among other things, is that our planet is heating up—for too much carbon dioxide prevents its normal loss of heat to the atmosphere. When our own bodies heat up in this way, we have a fever. If we don't solve our energy and production problems very soon in ways that are healthful for the Gaian body, the earth will have to solve the problem itself. Atmospheric carbon dioxide is rapidly approaching levels it apparently reached previously just before the ice ages. Perhaps Gaia will cool her man-made fever with a new ice age, destroying most of what we have built and forcing us into retreat, like the ancient bubbler bacteria, to safer environments.

Inconvenient as another ice age would be, we at least know humans have survived a number of them by moving to the tropics where new land is exposed as ocean water is removed to form snow and ice. Far worse would be Gaia's other alternative, which is now looking ever more likely—to reset her thermostat at a higher planetary temperature, thus regaining the Gaian system's stability at our expense, for we humans may not survive the increased heat.

This is what we are learning to understand—that the Gaian life system has evolved in such a way that it takes care of itself as a whole, and that we humans are only one part of it. Gaia goes on living, that is, while her various species come and go. We used to believe that we were put here to do whatever we wanted to with our planet, that we were in charge. Now we see that we are natural creatures which evolved inside a great life system. Whatever we do that is not good for life, the rest of the system will try to undo or balance in any way it can. That is why we *must* learn Gaia's dance and follow its rhythms and harmonies in our own lives.

5

The Dance of Life

It was in the search for life on other planets that we discovered what a live planet is—and that we ourselves are part of the only live planet in our solar system.

The first astronauts to see the whole earth with their own eyes were astonished by what they saw. Although they couldn't see any of the living creatures they knew to be on it, the earth itself looked very much alive—like a beautiful glowing creature pulsing or breathing beneath its swirling, veillike skin, as we saw it in our imaginary film.

Scientists, of course, cannot simply trust the way things *look*. After all, science was built on the discovery that the earth is *not* the unmoving center of the universe, much as it looks to be just that. Nevertheless, it was seeing our planet from afar for the first time and noting how very different from other planets it appeared that inspired new ideas and studies of Earth.

Long before we saw our planet in this new way, scientists had adopted the view that the earth with its various environments is a nonliving geological background for life, living creatures having evolved upon it by accident and having adapted to it by natural selection. The Scottish scientist James Hutton, who is remembered as the father of geology, seems to have been virtually ignored when, in 1785, he called the earth a living superorganism and said its proper study should be physiology. A century later the Russian philosopher Y. M. Korolenko told his nephew, Vladimir

Ivanovitch Vernadsky, that the earth was a live being, and though it is not clear that Vernadsky believed this himself, his studies of earth took a very different view of life than did those of other scientists.

Vernadsky called life "a disperse of rock," because he saw life as a chemical process transforming rock into highly active living matter and back, breaking it up, and moving it about in an endless cyclical process. The Vernadskian view is presented in this book as the concept of life as rock rearranging itself, packaging itself as cells, speeding its chemical changes with enzymes, turning cosmic radiation into its own forms of energy, transforming itself into ever evolving creatures and back into rock. This view of living matter as continuous with, and as a chemical transformation of, nonliving planetary matter is very different from the view of life developing *on* a nonliving planet and adapting to it.

While this Vernadskian view is stimulating much research in the Soviet Union, it never became widely known in the West. The biologist G. E. Hutchinson was one of the very few Western scientists of this century who took an interest in and promoted Vernadsky's view that life is a geochemical process of the earth.

Then the independent English scientist James Lovelock, at NASA during the search for life on Mars, knowing nothing of Vernadsky's work, shocked the world of science by suggesting that the geological environment is not only the product and remainder of past life but also an active creation of living things. Living organisms, said Lovelock, continually renew and regulate the chemical balance of air, seas, and soil in ways that ensure their continued existence. He called this idea—that life creates and maintains precise environmental conditions favorable to its existence—the Gaia hypothesis, at the suggestion of his Cornwall neighbor, the novelist William Golding.

The Gaia hypothesis is now recognized as Gaia theory, but it is still controversial among scientists. Lovelock, like his predecessor Hutton, calls Earth-as-Gaia an organism or superorganism and claims its proper study is physiology. Yet he also calls Gaia a self-stabilizing mechanism made of coupled living and nonliving parts—organisms and physical environments—which affect one another in ways that maintain Earth's relatively constant temperature and chemical balance within limits favorable to life. Lovelock describes this mechanical system as a *cybernetic device* working by means of feedback among its coupled parts to maintain Earth's stable conditions in the manner of a thermostat-controlled heating system that maintains house temperature, or an automatic pilot that keeps an airplane on course. This concept of Gaia as a cybernetic device is far more acceptable within the mechanical worldview that still dominates science than is the concept of Gaia as a live organism.

For Lovelock "organism" and "mechanism" are equally appropriate concepts, but in fact the two concepts contradict each other logically, and this causes confusion around the whole issue of Gaia theory. The concept of life—by any definition, including the autopoietic definition of self-producing and self-renewing living systems introduced in Chapter 3—is not logically consistent with the concept and reality of mechanism.

For one thing, life cannot be part of a living being; life is the essence or process of the whole living being. If Gaia is the living earth, then it would be as meaningless to say that life creates its own environments or conditions on earth as it would be to say that life creates its own environments or conditions in our bodies. Life *is* the process of bodies, not one of their parts, and in this book we maintain that the same is true for Gaia-Earth—that life is its process, its particular kind of working organization, not one of its parts. We can still say that organisms within Gaia create their environments and are created by them, in the sense that we say cells create their own environments and are created by them in our bodies. In other words, there is continual and mutually creative interaction between holons and their surrounding holarchies. But we do not divide living bodies or holarchies into "life" and "non-life."

If we accept the autopoietic definition of life, we see another contradiction between Gaia as a living being and Gaia as a mechanical system in which life and non-life are coupled parts. An autopoietic system is self-producing and self-maintaining. It must constantly change or renew itself in order to stay the same—your body renews most of its cells within each seven years of your life, for instance. No mechanism has ever done this, because a mechanism is not self-ruled but other-ruled—produced and repaired (or programmed for repair) from the outside. It cannot, and therefore does not, change itself by its own rules, and that one fact points out the essential difference between living systems and mechanical ones including even the most sophisticated computers and cybernetic robots. We will say more on this subject later, especially in Chapter 15 and in the Epilogue. For now, let us just note the contradiction that arises if we define Gaia at once as a living organism and as a cybernetic device—a contradiction that is causing confusion about Gaia theory among scientists.

The position of this book, then, is that the earth meets the biological definition of a living organism as a self-creating autopoietic system, and that only limited aspects of its function—never its essential self-organization—may be usefully modeled by cybernetic systems, just as we can usefully model aspects of our own physiology (for instance, temperature regulation) as cybernetic feedback systems.

Let us look now at the planet Earth as a self-producing living organism, which we call Gaia to distinguish it from a nonliving planet with life upon it. Lovelock's first clue to Gaia came to him when he was comparing the atmospheres of different planets. The atmospheres of the other planets in our solar system all make sense chemically—they are stable mixtures of gases. Only Earth has an atmosphere that is quite impossible by the laws of chemistry. Its gases should have burned each other up long ago!

Yet if they had, Earth would have no living creatures. And of course it *does*. They make and use almost the entire mixture of gases we call the atmosphere, ever feeding it new supplies as they use it and as it burns itself up chemically. This activity of living things always keeps the atmosphere in just the right balance for the life of Earth to continue. We can compare it to the activity of our cells in producing, using, and renewing the blood, lymph, and intercellular fluids flowing around them.

Living creatures, for example, produce four billion tons of new oxygen every year to make up for use and loss. They also make huge amounts of methane, which regulates the amount of oxygen in the air at any time, and they keep the air well diluted with harmless nitrogen. In fact, the Gaian atmosphere is held at very nearly 21 percent oxygen all the time. A little more and fires would start all over our planet, even in wet grass. A little less and we, along with all other air-breathing creatures, would die.

Every molecule of air you breathe, with the exception of trace amounts of inert gases such as argon and krypton, has actually been recently produced inside the cells of other living creatures. Thus the atmosphere is almost entirely the result of the constant production of gases by organisms. If these smaller organisms within the great Gaian organism stopped making and balancing the gases of our air, the atmosphere would burn itself up rather quickly. And if living things didn't turn salty nitrates into nitrogen and pump that nitrogen into the air, the seas would become too salty for life to go on in them, and the atmosphere would lose its balance. The right balance of chemicals and acid in the seas and in the soil, and even the balance of temperature all over the earth—all of the conditions necessary for the life of our planet, that is—are regulated within the planet as they are in our bodies.

Our sun has been growing larger and hotter ever since the earth was formed, yet the earth has kept a rather steady temperature—in much the same way that a warm-blooded animal keeps a steady temperature while things get cooler or hotter around it.

Old attempts to explain how geological mechanisms might regulate the earth's temperature are giving way to new explanations of how a live planet does it. Part of the complicated system involves regulating "greenhouse gases" such as carbon dioxide and methane, which trap solar heat; another part involves controlling the amount of cloud cover to let in more or less sunlight. Perhaps the earth even creates ice ages to cool its fevers.

In our own bodies, there are always things going on to upset the balance of oxygen or salt or acid in our blood and cells. Yet the parts of our living body work together constantly against these upsets of balance. Just so, it seems that the parts of the earth work together to help it recover from its own imbalances, though as yet we know little about how this is done.

Although we have learned much about how the complex coordinated systems of our own bodies function, we can hardly even dream of knowing everything involved in building and running such systems. We seldom reflect on the fact that our bodies work without asking anything of our conscious, thinking minds. We need not even be aware of what is going on, much less having to think or plan or *do* anything about it. And a good thing this is, because we would most certainly mess up our bodies' wonderful work if we interfered in it in an attempt to control it ourselves. Lewis Thomas, an American scientist who is best known for his popular essays on science, has said that for all his physiological knowledge, he would rather be put behind the controls of a jumbo jet than be put in charge of running his liver. Any one of our organs is more complicated by far than the most complicated computer we've invented—and it knows how to run itself, repair itself, and work in harmony with all other organs.

In some sense, even if not in the sense of conscious minds, our bodies and other living cells and bodies *know* what is good for them—they know just how they should be balanced as well as how to do the balancing. Physiologists call this still mysterious property of life "body wisdom." It is clear that such wisdom, or intelligence, can evolve without conscious mind or conscious purpose. So if we see our Gaian planet acting "intelligently" or "wisely" in its own interest, without planning ahead or being aware, without having anything like a mind in the human sense, it should not seem strange to us. We would do well to acknowledge and respect Gaian wisdom in the sense that Thomas suggests we respect the wisdom of our bodies.

The sooner we recognize and respect Gaia as an incredibly complex self-organized living being, the sooner we will gain enough humility to stop believing we know how to manage the earth. If we stay on our present course and cling to our present belief in our ability to control the earth while knowing so little about it, our disastrously unintelligent interference in its affairs will not kill the planet, as many people believe, but it *will* very likely kill us as a species.

Starting with physicists' current view of cosmic beginnings, we have seen that the universe has tremendous energy to spend—and that it spends this energy evolving itself into ever more complicated patterns, including those we recognize as alive. We have come to believe that the total useful, or working, energy of the universe—according to the laws of physics, in particular the law of entropy—is gradually running down. Yet living creatures collect, store, and increase working energy wherever they find it, violating this law. To keep the laws of physics consistent, scientists believe that in increasing energy locally living beings must be decreasing the energy of their environment at an even greater rate. Only thus would they satisfy the overall demands of the entropy law, otherwise known as the second law of thermodynamics—the law which says that things are running down as a whole. This implies that living things must use up and thereby degrade their environment, making it ever less useful to other living things.

On our planet this would mean that each form of life gradually uses up or degrades its environmental supplies until it chokes itself off and dies. Indeed it seems that some living creatures sometimes behave in just that way, as did the first bacteria, which used up the ready-made sugars and acids in their environment, and as we humans do when we use up and destroy our natural resources. But when one kind of organism creates such a crisis, the living Gaian system as a whole seems to find a solution. On a planetary scale we find the enrichment of species and their environments in variety and complexity—a single system recycling its supplies without running down the way mechanical systems do.

What about the planet as a whole living being, then? Does it degrade its environment as it organizes itself? Over billions of years—surely a more than adequate test for the law of entropy—our Gaian planet has continued to self-organize in ever greater complexity. It lives off the sun, to be sure, but the sun does not burn up faster because the earth uses its energy, and the waste heat given off by the earth cannot be construed as degrading its cold space environment. It would seem that the entropy law, which is one of the laws of thermodynamics, and was "discovered" to explain how certain nonliving mechanical systems such as steam engines work, can tell us nothing about living systems.

Again we run into a contradiction between mechanics and organics. Many non-scientists, many readers of this book, probably find it strange that scientists *do* try to explain life in mechanical terms, feeling intuitively (and rightly) that there is something wrong with the whole idea. In later chapters we will see how the mechanical worldview of science and society came about.

○ ○ ○

Recent discoveries in physics strongly suggest that the nature of the universe was from the beginning such that it would come alive however and wherever possible. Perhaps planets are to our galaxy something like seeds and eggs are to multicelled earth creatures in that far more of them are produced than can actually form new living beings. And perhaps, like the cells in our own bodies, the "cells" of the universe, in the form of star systems or planets, may be alive for a time and then die; after death, their components may be recycled—in other words, the energy locked up in their atoms and molecules may be used again by some other part coming alive and needing supplies to develop. Those parts of the universe that seem most lifeless to us may be something like its skeleton—providing a framework as does the core of the earth in supporting its living surface, or the deadwood forming most of a redwood tree under its living surface.

If we agree that nature is not mechanical but organic, why should we not understand the energetic motion of the very first whirling shapes in the early universe as the first stirrings of that self-organizing process leading to living organisms? The spiraling pregalactic clouds, composed of spiraling atoms, held themselves together, drew in more matter-energy from their surroundings, built it into themselves, and lost energy again to their surround. In this process of energy exchange they evolved into new, more complicated forms. By the time we get to galaxies and to fully formed stars within them, the dance toward life has become quite complicated already. We are only beginning to discover how complicated are the structure and process of our own sun star, and we still have much to learn about the way our own planet rearranges its matter into those lively chemical patterns we all agree to call living organisms.

Earth, we now know, is the only planet in our solar system that had just the right size, density, composition, fluidity of elements, and just the right distancing and balancing of energy with its sun star and satellite moon to come alive and stay so. Yet its life *is* a result of this fortunate confluence of conditions, just as the development of a plant or animal embryo is. Our living earth is likely no more a freak accident than is the seedling that grows or the frog egg that matures. All are the inevitable result of right compositions and conditions. Some scientists believe the conditions were so special that Earth is a rare phenomenon, perhaps the only such planet in the universe. But there is no better reason to believe this than there is to believe that living planets are as common in the universe as are the successful seedlings and hatchlings of Earth. And if this is so, there are billions of other live planets in the billions of galaxies, each with billions of star systems.

The only part of the earth that is more energetic than living creatures is the lava erupting or oozing through its crust, but most of that energy is quickly lost as heat pouring into the atmosphere, while living things recycle their energy within and among themselves and from one generation to another. On the whole, the living matter of the earth, as Vernadsky would call all its living creatures taken together, is up to a thousand times more active, more energetic, than the rocky crust from which it and they evolved. Hardly an example of the decreasing energy predicted by the entropy law! Where did all this energy come from?

The giant molecules from which the first creatures formed themselves were produced by powerful solar and lightning energy, some of which got locked up in them. The creatures formed from these molecules released this energy by breaking up other big molecules, or learned to use solar energy directly as we have seen, maintaining themselves and producing an oxygen-rich atmosphere in the process. Oxygen-burning respirers get their energy by consuming fermenters, photosynthesizers, and one another. Organisms can thus convert stored energy or direct solar energy into other useful forms of energy—the energy of motion, of heat, of chemical reaction, even of electricity—while the atmosphere regulates the kinds and amounts of solar radiation available, keeping it within appropriate bounds. As Lewis Thomas has said, earth seems to be a creature "marvelously skilled in handling the sun."

Meanwhile, the raw materials of the earth's interior spew or well up as new rock to be transformed into living matter, while old living matter, dead and compressed back into rock, sinks back into the soft mantle at the edges of tectonic plates. On the earth's surface scientists have a hard time finding any rock that has not been part of living organisms, that was not transformed into living matter before it became rock again. We will see examples of this process later.

Thus virtually all of the atmosphere and all of the rocks have been through at least one phase in which they were living matter. The same is true of the soil and the seas. It is easier to distinguish between life and death than between the domains of life and nonlife we have assigned to biologists and geologists. In fact, virtually every geological part or feature of earth we can find is a product of our planet's life activity. Further, living organisms have "invented" 99.9 percent of all the kinds of molecules we know, almost all of them back when bacteria were the only creatures around. This is autopoiesis—the self-production we take, in this book, as the definition of living beings.

○ ○ ○

What confused us for so long—kept us from seeing that our planet is alive as a whole—is at least in part our own human space (size) and time perspective. Since we easily see ourselves and many kinds of plants and animals as wholes separate from one another and from their surround, we have had as hard a time recognizing ourselves or them as parts of a single being as we had recognizing that we ourselves are made of separate cells. In one instance we saw the parts more easily than the whole; in the other we saw the whole more easily than the parts.

If we had a magnifying glass powerful enough to let us see everything in the world around us at the level of molecules, we would see life in the energetic molecular dance of chemical reactions and recombinations—the dance that weaves molecules into new patterns, some livelier than others. But instead, our experience comes through eyes that see life as a collection of separate plants and animals. This makes it hard for us to see them as parts of their environment, much less as parts of a whole living planet. Yet when we see the whole earth from far enough away to show it on a movie screen and speed up its rotations, it *does* look alive, though we can no longer see its "separate" plant and animal parts. We have no way of seeing our world of life-within-life at all its size levels at once, but we *can* use our minds to put information about different levels together and understand its living holarchy of holons.

The smaller living holons or organisms within Gaia grow and reproduce, so we have come to think of growth and reproduction as essential features of living beings. The autopoietic definition of life, however, does not include them as essential or defining features; rather, they are consequences of the autopoietic life process—something that may or may not happen, as when people do or do not reproduce. Therefore, the argument that the earth cannot be alive because it does not grow or reproduce does not hold.

Cells are the "packages" in which living matter housed itself when our planet came alive; they contain and connect autopoietic systems by enclosing them in open boundaries—membranes of their own making that allow materials and energy to be exchanged with the environment, as does the self-produced atmospheric membrane of the earth. In a sense, the whole earth is a giant cell within whose boundary membrane other smaller cells multiply, die, and are recycled in such a way that the whole need not grow. This is a wonderfully efficient way to make living beings (planets) possible in cosmic deserts with only stars for nourishment. Because our perception has been so focused on separate organisms in their physical or social environments, we tend to see insect, animal, and human societies, as well as whole ecosystems, as collections of individuals that have come to live and function together. It is actually more appropriate to see that such collections have always functioned as wholes which were never separated into completely individual beings. Individual creatures surely exist within and across species, but none could ever become completely independent, though some are relatively more or less independent than others. All their complex forms and ways have evolved within a single system, just as our cells evolved their separate functions within an inseparable whole. Their connections with their species fellows and with their ecosystem are as holons within holarchies up to the whole Gaian organism—connections that were never broken and cannot be, just as our cells cannot break their connections with their organs or their/our whole bodies.

Scientists who try to understand Gaia as a collection of separate organisms mechanically coupled to nonliving environments get bogged down in arguments about whether the organisms—collectively called the *biota,* or "life"—could actually have joined forces on purpose to "control" the conditions of their *abiotic,* or nonliving, environment in their own interest. How could all bacteria—assuming the Gaian mechanism was formed when there were no other creatures—get together, they ask, to work cooperatively and purposively for their own good?

This argument, like the other confusions previously discussed, is another result of seeing Gaia as a mechanism. Purpose is an essential aspect of machinery, all machines being built to serve the purposes of their inventors and users (more on this in Chapter 15). Thus, machines are "other-ruled," as we said earlier. And so, seeing Gaia as a mechanism raises the question of purpose. But purpose is a taboo in scientific descriptions of nature, because God, whom Renaissance scientists saw as the Grand Engineer of natural mechanism, is no longer part of scientific explanation. Scientists thus argue a logical contradiction: that nature is mechanical but has no creator and no purpose.

If we see natural organisms as self-ruled autopoietic systems that evolved without benefit of a purposive God, we see that they simply evolve wherever they are not prevented from so doing, wherever their energetic development is mutually consistent with whatever else is going on around them. No one argues about whether or not our bodies regulate our temperature "on purpose"—we simply accept that they do so because they evolved that way, "purpose" being limited to the relatively recent (in evolution) conscious mind, about which we will say more later.

The answer to the question of whether bacteria assembled purposively to control their environments is thus that they did *not* "get together" any more than did the cells of our bodies assemble themselves after they had formed. All bacteria are living matter transformed from earth's rocky crust and packaged in open boundaries that keep them functioning as a single system. They are not separate from one another or from the crust; they are not parts of an assembled mechanism but part of a single Gaian life process we can call *geobiological.* The Gaian organism has evolved to do what it needs to do in order to preserve itself as naturally as we do and with no more purpose than we find in our own bodies.

If Gaia is a single live planet or organism, why did its rock rearrange itself into such an astounding variety of individual creatures? Why not just a planet holon, instead of a planetary holarchy of holons?

We might as well ask why the first gas clouds sorted themselves into individual galaxies, and the galaxies into stars and planets and other space bodies. The answer, as we now begin to understand, is that life becomes ever more stable as it becomes more complex! Mechanical systems may be more vulnerable to breakdown as they become more complex, but this seems not to be true of living systems. The Gaian division of labor or function among different species—different *kinds* of creatures—makes possible a division of labor similar to that of our bodies, which function efficiently through the combined work of many different kinds of organs. No place, or environment, on earth—from the barest mountaintop to the deepest part of the sea—has fewer than a thousand different life species, mostly microbial, forming it and doing different things to keep it alive and evolving. If a planet does come alive, it would seem that it must come alive everywhere, not just in patches.

Scientists are only now beginning to work out the physiology of our Gaian planet—to understand why the introduction of a single new species into a complex environment can make that environment ill, just as the introduction of a single species of disease microbe into our bodies can make *us* ill; they are only now coming to understand why the destruction of an environment such as tropical forest can unbalance the whole planet, just as removing an organ from our bodies can unbalance us. Yet we are also discovering that Gaia's incredible complexity makes her tougher and more resourceful than we are. We are far more likely to choke our own species off by destroying our environment than we are to kill Gaia. Gaia's evolving dance of life will continue with or without us.

The word "evolution," when used in talking about human dancing, means the changing patterns of steps in any particular dance. A dance thus evolves when its step patterns change into new ones as the dance goes on. In exactly this sense, the evolution of Gaia's dance—of Earthlife—is the changing patterns of steps in the interwoven self-organization of creatures and their habitats over time.

We see that Gaia's dance is endlessly inventive. Trying out new step patterns in a dance is called *improvising*, as a creative dance is not planned out in advance. Rather the dancers improvise as they go, testing each new step for its fit with other steps and with the whole dance pattern. Gaia's dance seems to have evolved by such improvisation, the working out of basic steps used over and over in new combinations.

In Gaia's dance, all organisms smaller than Gaia, from the first bacteria to ourselves, have been built from DNA and protein molecules. The very complex patterns of these giant molecules are almost entirely made of only six kinds of atoms—hydrogen, carbon, nitrogen, oxygen, phosphorus, and sulfur. And as we saw, all kinds of atoms other than hydrogen were created all over the universe under pressure inside stars as combinations of the original hydrogen atoms, which were combinations of the original subatomic particles.

There are very few kinds of protein or other molecules on earth today whose patterns the ancient bacteria had not already invented billions of years ago. Nor have any new basic life processes been developed since bubblers, blue-greens, and breathers invented the three ways of making ATP energy molecules: fermentation, photosynthesis, and respiration. In other words, evolution since then has been a matter of rearranging not only the same atoms but also the same molecules and life processes into an endless variety of new creature patterns. This, then, is Gaia's dance—the endless improvisation and elaboration of elegantly simple steps into the awesomely beautiful and complex being of which we are the newest feature.

6

A Great Leap

O A century ago, priests told people that the earth was a few thousand years old; a few decades ago, scientists believed that life on earth began only slightly more than half a billion years ago. Now we know that the earth's skin was already swarming with fully evolved monera well over three and a half billion years ago. Two billion years—almost half the earth's life—belonged solely to the bacteria. This chapter is the story of the dramatic leap into the other four kingdoms of life, a leap that took place about one and a half billion years ago and which holds important lessons for humanity today.

Before we go into that story, however, let us recall the magnificent work of the bacteria in preparing the way for their own evolution into other forms of life. Let us recall that they invented all life's ways of making a living and created the conditions for life that we huge latecomers enjoy today. Without the bacteria, the earth's atmosphere would be unbreathable and its crust would have remained a cratered desert of glassy rocks. Without the activity of bacteria, even the oceans, as we saw, would have gassed off our planet.

When we left the story of evolution, the blue-greens were turning solar energy to use in making food, and turning food energy to use in the work of life. Their waste gas, oxygen, together with other waste gases, such as the methane made by the older bubbler bacteria, piled up, creating a new kind of atmosphere. Gaia had solved some big problems and was thriving.

The blue-greens tried out all sorts of new shapes for their one-celled bodies—tiny balls, big blobs, long strings of individuals joined end to end, and even great sheets of them all stuck together with jellylike stuff and looking like seaweed. Some, living in large colonies, branched out into shapes that plants adopted when they evolved much later; they even made spores in ways that remind us of plants making seeds. Perhaps the coded plans for such shapes and parts were stored in bacterial DNA and passed on for many millions of years until true plants evolved and found use for them. Other kinds of bacteria lived on their own as threadlike whips, lashing themselves from place to place. This was another early evolved structure that prefigured a way of moving that would evolve later in the dance.

Besides trying out new shapes and movements, the monera evolved a division of labor that streamlined individual bacteria by reducing the amount of DNA and equipment each had to carry. Various kinds of monera became specialists at particular jobs, such as respiration, photosynthesis, or fixing nitrogen gas, yet all of them had access to the whole bacterial gene pool because they never lost their ability to trade DNA when necessary.

The new, streamlined specialists spread out into new kinds of places. Some did well in freezing cold waters; others lived in very hot springs. Some blew about in the air they had helped to create, then settled far from where they began. Some even found it possible to live and multiply on land, eating their way into rocks, where they started the process that would eventually turn them into soil. But the more specialized the monera were, of course, the less independent they were, the more they depended on one another. Oxygen users now needed oxygen makers, food eaters needed food makers, nitrogen builders needed nitrogen fixers, and so on. In specializing, the monera were evolving what we call food chains, or ecological systems, in which each species provided for and took from others.

It seems that nature must always work out a balance between the *in*dependence and *inter*dependence of individual creatures—between their autonomy and their holonomy. Specialization—whether in human society, ecosystems, multicelled creatures, or bacterial networks—is a feature of whole systems that makes them more versatile and efficient through the interdependence it creates among parts. Specialization brings variety into the life dance, but increases holonomy at the expense of autonomy, since it increases interdependence. This balancing of autonomy and holonomy is very important to understand if we are to learn to manage our human affairs as well as Gaia has worked out hers.

We saw the bacteria multiplying to ever greater numbers, discovering ever more and different ways of surviving and making their living, adapting themselves to geological changes in their environment as well as to the changes they themselves brought about. Before the ozone layer had gathered, strong ultraviolet light often damaged their DNA, but they invented splicing enzymes to repair the damage, and they learned to share this information, as well as other information stored in DNA plans, with each other. Thus the gene plans—patterned bits of DNA—for many variations on their organization were kept available, to be borrowed and copied from one another when the need arose. The bacterial population as a whole could therefore respond to emergencies such as chemical changes in the environment by drawing on and quickly spreading the genes that best helped them cope with those emergencies. Nowadays, part of their coping in this manner is developing resistance to man-made antibiotics.

Huge teams of specialist bacteria were in effect forming the organs of the Gaian organism, performing its various tasks and recycling its materials, learning to balance the whole dance of life. But as specialists they also ran into trouble now and then. We can imagine that blue-green food makers—needing light, for instance—must have found themselves stuck sometimes in dark places, and bubbling or breathing food eaters must have found themselves short of food because there was so much competition for it, so many mouths to feed, so to speak.

Ultraviolet light reaching the earth had helped produce the sugars and other food molecules on which bubblers and breathers depended. As the ozone layer grew and began to screen out ultraviolet rays, this production was stopped, increasing the competition for food to the point of crisis. The challenge of worldwide hunger seems to have pushed the monera to rediscover some old steps and patterns in their dance of life and weave them together into a new pattern that produced a very great leap in the dance of evolution.

This leap was accomplished when the ever more specialized bacteria got together within the same walls, where they could use their various ways of making a living cooperatively. In doing so, they evolved a very large and sophisticated new kind of cell—a kind of cell so different from the bacterial moneran kind that it is more closely related to us than to any of its own ancestors.

○ ○ ○

The new cells—on the average a thousand times bigger than bacterial cells—formed a second kingdom of life to join the monera—a kingdom of life we call protista. The name comes from a longer word, *protoctists,* meaning "first builders." Protists, like monera, are single-celled creatures, though they will go on later in our story to build multicelled creatures, as they themselves are multicreatured cells.

Although the big new protists, once evolved, were smoothly run cooperative ventures, they did not start out that way. In fact, the new step in evolution almost certainly began quite uncooperatively in the desperate search for food. With the growth of bacterial populations and a developing ozone layer, the time when free food lay or floated all around came to an end. Natural death had not yet been invented to recycle materials, as bacteria do not necessarily or normally die and dissolve into reusable parts. Rather, parent bacteria split themselves to become their own offspring, and ever more offspring made ever greater demands on the food they needed in order to build themselves to full size.

In short, the earth's crust had come alive by packaging ever more of its atoms and molecules into bacteria, many of which depended on ready-made food supplies that were now limited because the activities of the bacteria had created a new atmosphere. The planetary process of coming alive was in danger of choking itself off by overcrowding and lack of food. It is not impossible that something like this happened to Mars—that Mars once came alive and then died for lack of supplies, or lack of a means to recycle supplies efficiently enough to keep its early creatures healthy.

On earth, the evolution of giant cell cooperatives probably began when tiny energetic breather bacteria began forcing their way through the walls of larger bubblers to get at their rich molecules—not entirely unlike the way in which we humans have invaded other kingdoms or countries to get at ready-made supplies and raw materials.

The problem with this approach was—as it still is—that eventually the invaders run out of supplies again, having eaten up their hosts. In the long run, invaders have little more to gain than their victims. But this crisis was a new challenge to life, and life proved as inventive in this situation as it had been in the face of previous problems.

Unable to get rid of, say, the invading breathers, the big bubblers seem to have negotiated an agreement with them that was of benefit to both parties. Perhaps in return for feeding on the bubblers' molecules, the breathers gave the bubblers some of the ATP energy they could make so much faster. This is not unlike the deals made between countries—when, for example, a rich country offers electrification in return for a Third World country's food products—and, as we know, it is not easy to make such deals truly beneficial to both sides. Such a bacterial agreement would have helped the big bubblers repair themselves or make extra molecules to feed the breathers in return for supplying them with plenty of energy.

Arrangements of this kind must have been worked out successfully, because the first protist fossils from about a billion and a half years ago show lots of breathers inside single big cells that were apparently healthy. This tells us that the breathers multiplied to large numbers within the host cell without damaging it. When such swelled cells divided, half the breathers went to one side, half to the other. And so the cooperative exchange of food for energy continued in ever more descendant cells.

Later such cells seem to have taken in still more partners—blue-green specialists in photosynthesis. The blue-greens were probably eaten by the big hungry bubblers, but they apparently resisted being digested. Eventually another cooperative agreement was reached: the blue-greens would make food molecules, which the bubblers and breathers needed, from water, carbon dioxide, and simple minerals plus sunlight. This cooperative arrangement allowed the giant cells to make ATP energy in all three ways within the same cell wall and thus to survive through all kinds of shortages.

As these new partnerships became more complicated, they also grew ever larger and heavier, no doubt finding it difficult to keep from sinking away from light or from moving too slowly to find food. Another cooperative solution—another transformation of imperialism into mutual aid—permitted them to move themselves farther and faster.

Remember that some bacteria were shaped like twisting, lashing whips? Others had invented a "proton motor"—a spinning disk that actually worked by electrical potential and had a tail attached to it so the bacterium could drive itself by propeller.

What might happen if these mobile bacteria stuck themselves onto larger cells in order to suck food from them? The big cells would find themselves moved about by lashing tails and propellers. This is just what seems to have happened. The evolving cooperatives must have found the pushy bacteria worth feeding in return for being driven to where they could find food or light. Eventually, the long, lashing whips and propellers evolved into flagella, and some big cells evolved them into shorter, stiffer cilia arranged in rows so their lashing or rotating movements could be timed like the oars of our ancient ships. Cilia were so successful that almost every plant and animal cell living today has some part or parts that evolved from these ancient rowing hairs.

But before these happy cooperatives had actually evolved, there must have been a phase in which it was not clear whether cooperation would win out over competition, when the evolving protists were like a factory of workers without management, or a world of separate nations trying to take advantage of one another and all trying to give orders at once. After all, each kind of moneron had its own DNA plans, its own welfare requirements.

What was called for was serious organization. And monera, as we have seen, had already evolved a very effective communications system that could make such organization possible. They "knew" how to trade DNA to revise their own plans; they "knew" how to work from a common gene pool. And apparently they drew on this experience to set up a new kind of gene pool, or information center. Instead of collecting one another's genes inside themselves, they streamlined themselves even further, giving up some of their DNA to a common gene pool of general cell plans, which became the cell nucleus. As time went on, the nucleus became a virtual library of information for producing proteins and, in ways we still understand poorly, probably took on the direction of the whole cell's affairs. The individual monerons of the new cell became less independent but more secure, more inseparable parts of the new wholes.

Nuclear DNA, originally donated by generations of a thousand or more members of each of the evolving cooperatives, continues even in modern cells to contain a tremendous amount of duplication. Many theories have been proposed to explain the repetitive nuclear genes, but perhaps there is no more reason for them than there is for all the duplicated documents in our own government offices. In any case, the huge quantity of nuclear DNA could hardly be kept in bacterial style as loose loops floating about in the cell. It would have gotten tangled and broken, messing up plans. Besides, the DNA might well have been destroyed by all the oxygen-making going on in these cells. And so it seems that the DNA was collected from each part of the cell, wrapped in protein, and stored within a protective nuclear wall as a kind of central cell office for keeping DNA plans in order and thus making possible better organization of all the cell's work.

This central office and library of a nucleus evolved into an amazingly complex but elegant organization, with various kinds of proteins giving it structural and management order as it got information from the rest of the cell and directed its processes accordingly.

When it was time for the cell to divide, each very long twisted DNA molecule split and replicated itself. The pair stayed buttoned together at one point along its length, by a *centromere*—meaning "central place"—also called a *kinetochore,* or "moving spot," while each member coiled itself around a protein core. This coil then coiled neatly around itself, that coil recoiled into a shorter one, and so on until it formed a compact *chromosome* thirty thousand times shorter than the five-inch DNA molecule was when it started out!

Later we will describe the next steps in cell division. Here we just want to indicate the remarkable feats performed in this *nucleus*—this DNA-protein information center that evolved as the most important new feature of the giant cell cooperatives. Every living being of earth, except the bacteria, evolved from these nucleated cells, meaning that every living being of every kingdom of life beyond monera is made of the same kind of nucleated cell. Biologists call this superkingdom of cells—which includes the protist, fungus, plant, and animal kingdoms—*eukaryotes* (pronounced "you-CARRY-oats" and meaning "with a *karyon,* or kernel, of a nucleus"). To keep things straight, we call bacterial monera *prokaryotes* ("pro-CARRY-oats"), which means "before a nucleus."

In evolving these protist cells, or bacterial cooperatives, Gaia's creatures rediscovered some of the independence they had had before they became specialists depending on one another. The giant eukaryote cells could now evolve new parts and ways of using them—all sorts of special membrane walls and other supporting structures, gas-bubble vacuoles to control floating and sinking, other structures and chemical systems that helped do new jobs. A means was even found to circulate the jellylike cytoplasm in which all these structures are embedded, providing a transportation system for supplies and wastes.

Eukaryote cells, as we said, are on the average a thousand times bigger than prokaryotes, with a thousand times more DNA. They are in many ways as complex as human cities. Until recently, the nucleus seemed most like an authoritarian government or dictatorship, containing all the information necessary and sending out orders for what is to be built, produced, carried about, or otherwise done. Now it seems increasingly that the governing of cells is more decentralized—that the whole cell governs itself, using the nucleus as an information resource center.

Biology has made more progress in understanding the detailed composition of living things than in understanding their organization as a whole, but this trend is now shifting. In any case, in our analogy with cities, some cell parts are structures like roads and buildings; others are chemical messengers, carrying instructions to and from the nucleus; some are production centers like factories; others perform services, taking in and delivering food, collecting waste, making repairs. With ever more powerful microscopes and moving picture microscopy, we begin to understand how busy and lively, how complicated and amazing, life is inside such cells. But what we see leaves much to be learned about how it is all organized to function so smoothly.

All the cells in our own bodies are eukaryote cells, but we are just beginning to understand their evolution from ancient bacterial cooperatives. The story of this evolution has been simplified here. In actual fact it is one of the most fascinating and difficult puzzles ever to challenge biologists.

Biologists, as we have said, are in the habit of studying living beings, or organisms, as though they are complex machinery we can take apart to see how they work. They got this idea from physicists who believed all nature to be built like machinery and thus to be describable in mathematical terms. Because physics, as the most basic and mathematical science, had status and power, its ideas were imposed on biologists, as we will see in more detail when we get to the story of human history in later chapters. In trying to understand cells as strictly mechanical systems, biologists have come to see nuclear DNA as a kind of central computer whose programs direct the activities of all the other mechanisms outside the nucleus.

Most biologists still believe that this process is one-way, like the dictatorship analogy, but more and more of them are coming to understand that nuclear DNA is reorganized in response to changes within and beyond the cell, that the entire cell, including its membrane or wall, is a creative autopoietic system.

○ ○ ○

The DNA plans and composition of our own cells are, of course, unique to humans and differ from those of frog and fern cells, bacterial or Bactrian camel cells. When it became possible to analyze DNA in detail, it also became possible to identify a species by its own particular DNA pattern. Many species look very much alike, yet can be distinguished by differences in their DNA patterns.

Among the parts of our cells outside their nuclei are large numbers of tiny things that produce ATP—the molecules that store energy. These cell parts have long been understood as little mechanisms for burning oxygen to produce the cell's energy. But with rising interest in and understanding of DNA, biologists made a very strange discovery: that these little machines have their own DNA, the coded plans of which are quite different from those of the nuclear DNA.

How could cell parts be of a different species than the creature made of those cells?

Clues soon turned up. However different this DNA is from the nuclear DNA, it was found to be rather like some other DNA that biologists knew about—the DNA of bacteria quite like the breathers that evolved billions of years ago! At the time this was discovered, the story of bacterial cooperation in the evolution of eukaryotes was still unknown.

The idea of *cell symbiosis*—the origin of eukaryotes as prokaryotes living together in cooperatives—had been proposed independently by a German, an American, and a Russian biologist around the turn of the century. All had noticed that the photosynthesizing *chloroplasts*—meaning "green producers"—in the cells of plants resembled blue-green bacteria. The Russian, K. S. Mereschovsky, suggested that other ancient bacteria had evolved into other cell parts. But biologists, who were trained to see living things as put together from mechanical parts, could not see cell parts as creatures in themselves.

Thus the symbiosis theory was ignored until Lynn Margulis, an American microbiologist who became James Lovelock's partner in developing the Gaia hypothesis, revived it and produced a great deal of evidence to support it.

After much work, Margulis and others have shown that these energy-producing cell parts really are descendants of the ancient breather bacteria that came to live inside larger prokaryote cells, cooperating in building the first eukaryote cells—the protists, or "first builders," of all the kingdoms of life except monera, which had built *them*.

Luckily, teams of biologists working to unravel the ancient mysteries of cell symbiosis have found many clues in the behavior of today's bacteria. Rather vicious breathers can still be found drilling their way into other bacteria to reproduce there and eat the host bacteria from the inside. In the Tennessee laboratory of Kwang Jeon, protist hosts so invaded learned to tolerate and then to cooperate with their invaders in a mutually dependent relationship that brought about a new kind of creature. Surprisingly, this replay of the ancient evolutionary shift from outright aggression to full cooperation happened in only a few years' time.

Today, we find the descendants of the ancient breathers living and multiplying in the cells of every kind of protist, fungus, plant, and animal. It's high time we knew them by name. They are *mitochondria*—"mite-o-KON-dria," a word that comes from the Greek meaning "thread grains," because under a microscope they look like tiny grain hulls packed full of thread.

Using the oxygen we breathe, mitochondria make all the energy our bodies need to keep going and to repair themselves. Without our mitochondria we could not lift a finger. In fact, it is these swarms of bacteria, working night and day in all our cells, that keep us alive.

Or are *we* working for *them?* Lewis Thomas, in his popular biology essays, has suggested that if anything in nature is a machine, perhaps it is *us*. Maybe we are giant taxis which mitochondria built to travel around in safely and comfortably. Certainly mitochondria have done very well spreading themselves all over this planet, inside every other living thing, almost since the earth came alive. There are so many of them in our own cells that it's hard to guess at their actual numbers, but all together they would weigh almost as much as the bodies they live in—that is, mitochondria make up much of our weight, and the weight of elephants and insects, clams and monkeys, toadstools and lizards, fish and worms.

In plants—from seaweed to sunflowers, potatoes, and palm trees—mitochondria live together with their relatives, the chloroplasts, which give plants their green color. You will easily recognize them as descendants of the ancient blue-greens and know that as they make energy from sunlight, water, and carbon dioxide, they also make the oxygen their mitochondrian cousins need for making *their* energy.

Mitochondria and chloroplasts, together with their still free-living monera cousins, the bacteria, are by far Gaia's most numerous and important creatures, though we are very late in recognizing our complete dependence on them. Quite to the contrary, we discovered bacteria in the context of medicine and treated them only as enemies. They are so hidden and tiny that, for years, we paid no attention to their good works, but we now know that their cooperation is the essence of the entire Gaian life system. Besides making the vital gases oxygen and carbon dioxide for each other, the chloroplasts and mitochondria of eukaryotes, together with their prokaryote cousins, form other cooperative cycles—for example, the food chains mentioned earlier. While plants make their organic bodies from simple minerals, water, and carbon dioxide, animals can only make their bodies from the ready-made organic molecules of plants and other creatures. Bacteria, in turn, cause the decay of dead plants and animals, reducing them to the simple substances on which new plants can live.

Margulis's discovery that eukaryote protists evolved by cooperating in order to overcome the problems caused by competition among prokaryote bacteria was almost as much a shock to the world of science as was the Gaia hypothesis itself. Besides showing that cell "mechanisms" are creatures in their own right, she was suggesting that cooperation played a big role in evolution, while ever since Darwin, scientists—in Western countries, at least—had believed that evolution was driven only by competition.

The theory of evolution through competition has played a big role in our world, not only in science but in shaping our whole human outlook and way of life. Only by understanding its origin and its widespread effect on our lives can we understand how to change our view of ourselves and our role within our larger Gaian planet. Let's look, then, into the way our Darwinian view of evolution came about.

7

Evidence of Evolution

Charles Darwin was an English gentleman scientist who lived during most of the nineteenth century and who traveled far and wide to study nature. Seeing the great variety of plants and animals around the world, he was struck by the way each kind was suited to the place in which it lived. He believed, as had a few other scientists, that all these living things must have changed over time to fit their environments so well. This was a novel idea at a time when almost everyone believed that the world itself was unchanging and that God had created all species at the same time, just as they are. How could they have changed? Despite Darwin's observation of likely changes, he had no theory, no way to explain how they could occur.

Nature, he noticed, produced great numbers of seeds and eggs for all kinds of plants and animals, though only few of each kind grew up. Somehow it seemed that only those best suited to their environment—the "fittest" of each generation—survived to reproduce. But how could nature recognize and choose them?

Darwin had seen plant and animal breeders choose the fattest grains or the fastest horses of each generation to be the parents of the next generation. This was possible because, in each species, the young were not all exactly alike, but were as varied as human brothers and sisters. By such selection, generation after generation, the breeders changed the species, producing ever fatter grains or faster horses. This was exactly the kind of thing nature seemed to be doing, but how could nature select the fittest creatures of each generation?

The theory of evolution finally came to him when he read an article about food shortages and starvation. Could it be that all nature's young had to compete for food when there wasn't enough for all? Was this how nature put creatures to the test? If so, then surely the fittest in this competition would pass the test and survive to grow up.

The fittest bears, for instance, would be the ones with heavy coats to keep them warm as they hunted for food in cold places; in warm places bears with lighter coats would survive. Birds with long beaks were the fittest where worms had to be pulled out of holes; birds with short, strong beaks would survive where seeds had to be cracked. The bears or birds born with the fittest coats or beaks would win the struggle for food and grow up to produce babies while others starved or froze or were eaten. Their babies would inherit more or less fit coats or beaks, and again the fittest of all would be selected in competition.

Everything seemed clear now. Large numbers in the face of too little food produced competition, and competition led to *natural selection*. The selection itself was based on the natural variety of creatures, some of which were fitter than others.

In Darwin's time, of course, environments—such as deserts or mountaintops or sea bottoms—were seen as places in which living things made their living, not as live parts of a great living planet. Each environment would select for different kinds of fitness in the competition for food, for mates, for the best hiding or nesting places, and so on.

If the wind scattered the seeds of the same plant into different places, for instance, a desert would select for young plants that could live on the least water, while a windy mountaintop would select for those with the strongest grip on rocks. Just so, some thick-coated bears that could do well in cold environments might wander to warmer places and die out while bears with thinner coats survived.

Offspring of the same original parents might therefore become quite different over many generations as a result of settling in different environments that selected for different body patterns or features. When they became so different that they could no longer mate with each other, they had become different species. Thus Darwin explained the evolution of new species.

In Darwin's theory, then, unexplained accidents of birth that made creatures fit better into their environment were selected for survival and passed on to future generations. Unlucky accidents that made creatures less fit were rejected by natural selection and died out.

Scientists quickly made Darwinian evolution fit the idea of nature-as-mechanism by regarding creatures more or less as wheels fitting into the cogs of other wheels in the great clockwork of nature. Some wheels just happened to be made better than others by lucky mechanical accidents during their replacement, or reproduction. The idea of natural competition leading to the survival of the fittest appealed to men who were obsessed with the new social structure of capitalism.

These accidents of birth were eventually discovered to be changes in genes, then in DNA itself, and were believed to be pure, or random, accidents—"random" meaning without any pattern—that occurred as DNA copied itself or was damaged. Most biologists today still see accidents, now known to occur in DNA, as the only source of natural variation.

Ever since Darwin our general view of evolution has been of a battle among individual creatures pitted against one another in competition for inadequate food supplies. Only now are we in a position to understand the whole earth as a living body—a single dance woven of many changing dancers and their complex patterns of interaction. Competition and cooperation can both be seen within and among species as together they improvise and evolve, unbalance and rebalance the dance. Evolution *is* this improvised dance in which ecological balance is worked out over and over. Remember that living things have to change in order to stay the same—they have to renew themselves and to adjust to the changes around them. Rabbits evolve together with their "rhabitats," so to speak—all creatures evolving in connection with all else evolving around them.

It took a century and more after Darwin's theory was published for us to understand that environments are not ready-made places that force their inhabitants to adapt to them, but ecosystems created by living things for living things, even if not intentionally. All the living things belonging to an ecosystem, from tiny bacteria to the largest plants and animals, are constantly at work balancing their lives with one another as they transform and recycle the materials of the earth's crust.

Darwin, Lamarck, and Wallace all showed us that species actually do evolve over time, and attempted explanations of how this could happen. Their theories were a great step forward for science, since religion had put an end to all theorizing about evolution for nearly two millennia, since the most ancient Greek philosophers, such as Anaximander, had thought about it. But the mechanical framework of Darwinism—and more so the updated version known as neo-Darwinism—is a misleading way of seeing nature. The notion of the separateness of each creature, competing with others in its struggle *against* nature's challenge, fit society's own change to competitive and exploitative industrial production. Now, as we ourselves must learn to harmonize our ways with those of the rest of nature instead of exploiting it and one another ruthlessly, that notion makes little more sense than the notion that our own cells are separate beings competing with one another to survive in our hostile bodies. It is no longer useful or productive to see ourselves as forced to compete with one another to survive in a hostile society, surrounded by hostile nature.

The point here is that we *do* see ourselves in such competition, not because this is natural or sensible, but because the European or Western tradition of science has developed in close harmony with its social and political traditions. The Darwinian theory of evolution was applied to forming a society, a social system designed in accord with, and justified by, the Darwinian concept of nature. If we learn to see evolution as a single holarchy of holons working out cooperative health and opportunity over time, we can set up a social system to match *that* view. History may someday record the greatest discovery of twentieth-century science not as nuclear power or electronics but as the recognition that there is no absolute truth to be discovered about the world—that scientific theories can be judged only by their usefulness to science and ultimately to all society. Such usefulness changes over time, and thus scientific "truths" must necessarily evolve along with human society.

Neo-Darwinism insists that random accident and natural selection are the sole "mechanisms" of evolution. Yet much evidence is accumulating that well-organized, healthy creatures and their ecosystems are produced by self-organization, such as that which we saw in the genetic trade of bacteria and in their negotiated organization of nucleated cells—processes that can hardly occur by the simple accumulation of lucky accidents. Nor does natural selection adequately explain evolution, since it tells us little more than that some creatures die before they reach the age of reproduction. A modern theory of evolution must concern itself with the way in which natural holons are organized and maintained in holarchies, with descriptions of continual interactions among the levels of DNA, organisms and whole ecosystems.

As pointed out earlier, no place on earth today, not even the barest-looking mountaintop or the deepest part of the sea, has fewer than a thousand different species from various kingdoms—monera, protists, fungi, plants, and animals. Yet what we humans see as living things are only the largest of the plants and animals—beings the size of bugs and bushes and beluga whales, creatures on our own size scale. The vast majority of earth creatures, however, continue to be microscopic monera and protists. Think once again of our rocky planet rearranging itself through chemical activity into a rich network of bacteria and environments that are good homes for bacteria. This is what most of the activity of our living planet is still all about.

As Gaian evolution continued, some parts of each ecosystem went on rearranging themselves, as we will see, into ever larger living bodies grown from single cells: worms and insects, fishes and amphibians, funguses, flowers and trees, reptiles, birds, and mammals. Still, the smallest living creatures are even now those that work hardest to create the environments needed to sustain the larger plants and animals, as we saw in the last chapter. These larger creatures, for that matter, are really extensions of the microcosm. We have already seen that the eukaryote cells of which they and we are made are actually microbe cooperatives. And what are larger creatures if not the collective descendants—in fact, the *clones*—of such cell cooperatives?

In every Gaian ecosystem, we can regard each creature as something like a cell in our bodies, each species something like an organ. Each ecosystem is an ecologically balanced body evolving as a whole. We really should talk about co-evolution, rather than evolution, to remind ourselves that no species can or does evolve by itself, but that all must cooperate by adapting to the others' steps in the dance of life—by seeking, as we put it in earlier chapters, their mutual consistency with one another and with the rest of their surround. The story of co-evolution is still being put together as we try to understand that creatures are not just passive mechanical cogs in a wheel but active agents in their own evolution, or co-evolutionary process—the interacting autopoietic systems that compose the liveliest elements of planetary autopoiesis.

To some extent we can study the co-evolution of species within their present ecosystems and find clues to how they must have co-evolved in the dim past, but we also have many other clues to life's long-ago patterns. Fossils—the prints and remains of creatures turned to rock—are very important clues to what early creatures looked like, especially now that we know how to tell their age and have found fossils of even the most ancient bacteria in rocks, such as the stromatolites they built billions of years ago. Fossils alone cannot prove that one species changed into another, but the fossil record clearly shows that larger, more complicated creatures existed only after smaller, simpler ones had paved the way for their existence.

In Darwin's theory, species change very slowly but steadily as environments select for the tiny accidental changes that make some individuals of each generation slightly fitter than others. Since his time, however, many more fossils have been collected, and we see that most species must have changed in spurts from time to time—far more quickly than such accidents can explain—while others have scarcely changed at all, despite the slow, steady stream of accidents they must have endured. It is difficult to argue for a relationship between the rate of DNA accident and the rate of change in creature patterns. Some bacteria living now are like those that lived billions of years ago. Squid and sharks and ants have stayed much as they were hundreds of millions of years ago, as we see clearly in the fossil record. Yet many of these slow-to-change creatures' fellow species in co-evolution have become very different modern species.

Species whose genetic plans have hardly changed are like bicycles in a world of jet planes—they still work very well as they are, yet they have been steps along the way to bigger, more complicated inventions. They are, in a way, living fossils to compare with creatures that apparently continued to change or evolve. It seems that co-evolution has rhythms like any other dance—some slow, some so fast in comparison that they seem almost to leap from being one kind of creature to being another. We ourselves are a good example of a species that changed very rapidly.

If all living beings were created at once—as creationists still believe—then all modern species should have fossil ancestors quite like themselves. But this turns out to be true of only very few creatures larger than monera. There are estimated to be somewhere between three million and ten million species alive today, yet well over 99 percent of all the species that ever lived are now extinct. It is worth noting here that in the times of most rapid extinction it is estimated that the rate was about one species lost every thousand years, while we humans are causing the extinction of up to one million species in the last quarter of this century alone!

Now and then fossils give us wonderful clues to change, such as those showing reptiles evolving into birds. The archaeopteryx—which means "ancient wings"—was a kind of dinosaur with a horny beak and wings strong enough to propel its big body through the air. Its skeleton looks very much like those of modern birds, and the fossil imprints are so clear we can almost see reptile scales evolving into feathers, front legs into wings, and long snouts into hard beaks. The babies of birds, even today, still hatch from eggs just as their reptile ancestors did.

The fossil and chemical geological records show that dinosaurs and other large creatures of their time died out around sixty million years ago during a period when there were major changes in their environment. Apparently these large creatures could not adjust to big changes in climate caused, it now seems, by the impact of a giant meteor or planetoid. Smaller animals did survive the crisis, but they must have evolved so quickly that none of them left fossil records of the gradual changes inferred by Darwinian evolution.

Such jumps in the fossil record are too big for fossils alone to prove modern creatures' lines of descent. For that matter, scientists have not found a single clear and complete fossil line of descent for any modern creature, and this has become a major argument used by the creationists, who don't believe in evolution. How *can* we be so sure that modern creatures actually descended from earlier ones that were quite unlike them?

We have, fortunately, other clues to evolution. For one thing, we can actually *see* evolutionary changes taking place in some living species. In bacteria, protists, fungi, plants, and animals that reproduce quickly and often, we can follow the changes over many generations. With electron microscopes and other modern instruments we can actually track their changing patterns of DNA as well as the more obvious changes in cell or body structure that follow from the microscopic changes.

It was of course a big surprise for biologists to find that creatures can rearrange their own DNA in ways that can hardly be called accidental. In using antibiotic drugs to kill particular kinds of bacteria, as mentioned earlier, we often find that a species we attacked successfully has suddenly changed into a species that cannot be harmed by the drug. We say it has become *resistant*.

But such new resistance implies genetic change—the kind of genetic change we now know all bacteria to be capable of through the DNA recombination system that is as ancient as they are—the system that helped them cope billions of years ago with such life-threatening matters as ultraviolet radiation and poisonous oxygen in their environment.

Even more impressive is the fact that larger, more complex creatures also can change their DNA in emergencies. Many insects and plants that we humans consider pests have made themselves resistant to our chemical poisons in just a few generations. Recent work in biology, such as that of Mae-Wan Ho, shows that instructions do not flow only one way from nuclear DNA to bodies, but that bodies can register changes in the environment and in themselves, communicating these changes to the DNA in ways that offspring may inherit.

Just as we are finding that organisms are not separate from their environment, we are also finding that DNA is not separate from *its* environment within and without its organism. We also know now that the DNA recombination capabilities of cells include the ability to reproduce DNA between cell divisions—that genes can be copied and relocated when necessary. And this is only the beginning of our understanding of how organisms change their plans in order to survive in health.

The idea that the environment can produce inheritable changes in creatures had actually been proposed before Darwin's opposing idea that environments can only select among changes produced independently of environments. Jean Lamarck, who named the study of living things "biology," proposed it as the first modern theory of evolution. Lamarck's explanation of how living things changed themselves—the classic example being giraffes, which stretch their necks to reach higher leaves and then pass on long necks to their offspring—was not as convincing to Western scientists as Darwin's theory of accidental variety and natural selection through competition. Thus Lamarck's theory was ridiculed while Darwin's was adopted in the West, though Lamarck continued to be respected in Russia.

Neither Lamarck nor Darwin, of course, knew about DNA or even had a theory of genes, so they could not even guess that bacteria can trade around their genetic material or that larger multicelled creatures can rearrange their genetic material on the basis of experience in their environment. For that matter, most biologists still resist accepting the new Lamarckian evidence. We still have a great deal to learn about biological information systems. Earlier we mentioned that nuclear DNA contains a greater variety of genes than are needed to account for body organization, as well as many extra copies of the same genes. We guessed that the excess may be a hangover from the time when many bacteria contributed to the nuclear DNA. But it may also be, as some biologists believe, that all through evolution species collect and pass on reserve genes which may someday be useful in an emergency.

It even seems ever more likely that DNA can reorganize itself—or be reorganized by other cell components—to repair the kind of accidental change that was thought to be the only way to evolution. It would be nice to think we are not just a lot of piled-up accidents and copying mistakes, but beings who have at least in part organized and evolved ourselves in harmony with the other living beings that form our environment.

Certainly the evidence is mounting that co-evolution in ecosystems allows creatures to change themselves from the inside in response to their environment and to prod changes in other creatures as part of their environment. A change in one species will be reflected by changes in others. The evolution of whole ecosystems takes place as living things evolve themselves and are evolved by one another, as they incorporate raw materials into their bodies and are transformed in turn into raw materials for others. Environments, as we said, are not fixed places that living things must fit into, adapt to, as we had thought, but the busy activity of living things themselves, working out their ways of life together as parts of the live earth.

We began studying ecology only a few decades ago when we recognized things going wrong in our environment because of changes we were making—creating erosion and deserts by cutting down forests, poisoning land and sea creatures all over the planet with our pesticides, destroying our air and water and soil supplies with other polluting chemicals, and so on. We began to study our planet's ecologically balanced body, that is, with attention to ecological "illnesses" and their possible cures, just as we had begun earlier to study our own bodies with attention to what went wrong with them—in other words, to their illnesses. But as Lovelock has pointed out, medicine began to make real progress only when we began studying the *physiology* of our bodies—the way their interwoven systems work under normal, healthful conditions—and so we will make more progress in understanding ecology and evolution as we learn how our planet's normal physiology works. How are its many kinds of supplies recycled? How do its information systems work to adjust imbalances? How do its countless and varied creatures contribute to its overall health?

The more answers we find to questions such as these, the better we will understand how the Gaian system works now, and the more progress we will also make in understanding how it evolved.

○ ○ ○

Much groundwork for planetary physiology can be found in the work of V. I. Vernadsky. Vernadsky's concept of life as "a disperse of rock," paraphrased in this book as "rock rearranging itself," is very different from our usual view of life as a collection of individual creatures in various degrees of competition or cooperation as they adapt to a nonliving environment. Vernadsky pointed out that living organisms were originally built of the "inorganic" minerals of the earth's crust, that they still contain such inorganic minerals, and that they transform inorganic minerals into living matter and living matter back into inorganic minerals. For this reason he saw no separation between biology and geology, but became interested in the constant transformation going on from the one domain to the other. His concept of all living creatures together as "living matter" is one that postulates that part of the earth's crust is currently energetic enough to actively transform the more passive parts into itself and its products.

On the surface, this concept of living matter is the same as Lovelock's concept of "biota"—the sum total of living creatures, contrasted with the "abiotic," or nonliving, environment. But in Vernadsky's conception the emphasis is on geological continuity, on each as a transformation of the other, whereas in Lovelock's conception the emphasis is on their interaction as separate parts of a working system. Oddly, Vernadsky, who apparently did not see the planet alive as a whole, perceived its integrity more fundamentally than Lovelock, who does see it as an organism.

The processes by which organisms build up and destroy their own protoplasm—the mixture of proteins, water, carbon compounds, and minerals of which all organisms are made—is called *metabolism,* from the Greek word for "change." Metabolism is a process of chemical changes in living matter by which energy is provided for taking in new matter, building and repairing cells, collecting and excreting wastes. Metabolism is divided into two parts—*anabolism* and *catabolism,* the buildup and breakdown of protoplasm.

Metabolism, then, is the most basic (autopoietic) activity of all life—recycling the materials of the earth's crust into living matter and then back into nonliving matter that can be used again to create more living matter. Vernadsky understood metabolism as the activity of all living matter taken together as well as that of any particular organism, since he saw all living matter as a constantly shifting high-energy portion of the earth's crust. Since, as we said before, virtually all of the earth's atmosphere, seas, soil, and rock, even to its purest, hardest diamonds, is made from the products and dead bodies of organisms, it is clear that life is the most powerful of "geological" forces. The record of Earthlife's evolution lies in all of geology, not only in recognized fossils—a record referred to in the title of a book on Vernadsky's work, *Traces of Bygone Biospheres.*

The sedimentary rock formed by pressure on the ocean floor, for example, begins as sediment, including vast quantities of algae and animal shells, all passed through the guts of sand- and mud-eating worms to further transform them, just as soil is transformed by the related earth-eating earthworms of dry land.

The geological activity of creatures also includes their production of atmospheric gases and their transfer of groundwater back into the atmosphere, a process that is clearly visible in the pumping action of rain forests—the rain then falling to dissolve more earth and rock. On the whole, however, the geological activity of creatures is less the larger they are, most of this work being done by microbes and rock- or mud- or earth-eating worms. Some microorganisms contain half a million to a million times as much of some mineral, such as iron, manganese, or silver, as their environment does. The concentration of elements is one way in which life moves the earth's crust. Microorganisms are even responsible for concentrating the radioactive materials, such as uranium, that we mine to produce atomic energy—probably just to keep themselves warm!

It was so clear to Vernadsky that the activity of living matter was metabolic that he proposed we reclassify living organisms on the basis of their metabolism. He argued that our present classification from kingdom to species by way of phylum, class, order, family, and genus had led us to classify as related organisms many that really are not related under natural conditions. A better scheme, he felt, would be to divide kingdoms according to the way in which each of their species metabolizes supplies from its environment. The different ways in which organisms feed themselves had already been named by the German biologist Wilhelm

Pfeffer. Vernadsky proposed them as a biological classification scheme for the evolutionary geochemical, or metabolic, process of living matter.

In this scheme the metabolic process of organisms begins with the category called *autotrophs*—"self-feeding" organisms that can build their own giant molecules, such as protein and nucleic acids, from simple molecules and elements such as minerals, water, and carbon dioxide. This category includes the *photoautotropic* self-feeders that use sunlight in metabolizing basic molecules. A second major category of organisms is called *heterotrophs*—"feeding off others"—because its members cannot make large molecules from basic ones but must eat other organisms for their ready-made large molecules. A third category is called *saprotrophs*—"feeding on the dead"—because its members eat dead bodies and reduce their large molecules back to the basic ones the autotrophs can use. The fourth category is *mixotrophs,* which can metabolize in more than one way.

Finer distinctions within these categories are made as heterotrophs feed off other heterotrophs, and so on. What is important about this scheme is that organisms are classified not by their structures but by their functions within the whole geobiological life process. It recognizes organisms as self-organizing packets of the earth's crust with enough energy to move about the more sluggish matter around them. Vernadsky even suggested that evolution may proceed by the natural selection of organisms which most increase *biogenic,* life-originated, energy—the energy to move around the atoms and molecules of the earth's crust at the highest speed.

The energy of living matter sometimes explodes almost beyond belief. A locust plague of a single day has been estimated to fill six thousand cubic kilometers of space and weigh forty-five million tons! It is the locusts' heterotrophic metabolism, of course, that makes them a plague as they suddenly convert vast quantities of the autotrophic crops, planted by humans, into their bodies. Most biogeologic activity goes on less dramatically—though it is impressive enough to consider that a single caterpillar may eat two hundred times its weight per day. All ecological areas have more autotrophs than heterotrophs—it takes more of the former to sustain the latter. Thus, a forest may have 2,000 to 5,000 times as much autotrophic as heterotrophic living matter.

Vernadsky did not consider this classification scheme the only one possible; he recognized that one can learn much by trying other methods. One of his schemes was based on the type and amount of mineral content in organisms. Certainly he was one of the first modern scientists to see the earth in a truly holistic way and to provide evidence of its evolution through the transformation of rock into living creatures and back into rock.

Many scientists have since built on his work or developed independent studies of the earth from a holistic perspective, but Vernadsky's work has been given particular attention here because its fundamental conception of biogeochemical unity is so important and so little known in the West. Recent and more easily available books—Lovelock, *The Ages of Gaia;* Margulis, *Early Life;* and Margulis and Sagan, *Microcosmos*—present an updated account of the earth's evolution as a planet, as Gaia.

8

From Protists to Polyps

If you look at a drop of pond water under a microscope, you will see a great variety of protist creatures living together, creatures you would never have suspected were there. You may see, for instance, paramecia, tiny slipper-shaped algae rowed along by hundreds of waving cilia. Then along will come a giant, blobby amoeba, changing shape before your eyes as it pushes out pseudopods, or "false feet," in its search for food. Perhaps it will get stuck in a tangle of rodlike algae, strings of cells sitting quite still and making energy from light with their green chloroplasts. Other green algae, such as euglena with their long whiplike tails, may flit by.

If you look instead at a drop of seawater, you may see whirling dinoflagellates that glow in the dark by making their own light, or diatoms and radiolaria, like incredibly beautiful crowns or glass ornaments. There seems to be no end to the fantastic patterns of these tiny single-celled creatures, whether they live alone or stuck together in colonies like the ball-shaped volvox.

The world of single-celled creatures in a drop of water is probably much as it was a billion years ago when there were no larger creatures. Yet these same protists went on to build and evolve all the larger multicelled creatures of earth.

On today's oceans great blankets of plankton—mostly made of protists—float about ever renewing the atmosphere with nitrogen, fresh oxygen, methane, and other gases they release. They also provide food for many heterotrophic species, from the almost microscopic shrimp that swim among them to the largest of whales. The wastes and dead bodies of those that eat plankton sink to the bottom where myriad saprotrophs cause them to decay back into molecules that may come to the surface to nourish new plankton.

Plankton not only serve as part of a food cycle but play a most important role in balancing the chemistry of the atmosphere and the seas as well, and in the biogeology we talked about in the last chapter—the transformation of the earth's crust from rock to living matter and back to rock.

Among the creatures forming plankton are thousands of species of diatoms and radiolaria—those beautiful protists whose shells look like fancy glass ornaments when seen through a microscope. In Gaia's dance there are more diatoms than any other kind of creature except bacteria. And they are responsible for moving a great deal of rock about—helping to transform rock into sediment in the dance of life.

Land rock, we know, is dissolved by streams and rivers into salts and minerals, which they carry to the sea. Life in the sea needs these building materials, but like the gases of the air, they must be kept in balance. If too many of them pile up in the sea, living creatures will choke.

One such mineral is silica. A huge amount of silica is washed into the seas every year—hundreds of millions of tons of it. But huge numbers of diatoms wait in the sea for these silica supplies. Diatoms use a great deal of silica to build their fantastic sparkling shells. When they die, the diatoms sink to the bottom, leaving their silica shells to settle into rock—three hundred million tons of silica rock every year!

The number of diatoms in the sea naturally adjusts itself to the supply of silica brought by the rivers. There is always just the right number of diatoms to use up the silica dissolved in the sea. This well-balanced system—water dissolving rock, diatoms making it into lovely silica shells, then being eaten by others or dying so their shells sink down to the sea bottom to become new rock—worked itself out in co-evolution as a kind of transport system within the great Gaian body. The shells of radiolaria are also part of this process. The famous white cliffs of Dover on the English coast are ancient deposits of the microscopic, chalky, snaillike shells of another marine protist—foraminifera—which were pushed out of the sea by the endless motions of the earth's crust.

Many other biogeology cycles or systems are still to be discovered as we keep working on the puzzle of planet physiology.

Certainly it is very different to see creatures as living parts of the great Gaian body than to see them as we used to, as individual creatures struggling to survive in an alien environment.

Unfortunately, our old view of nature made us see ourselves as just one of the many creatures competing for survival on a planet without enough for all to live. But now that we have microscopes, telescopes, spaceships, computers, and other instruments that show us so much more than we could see with our eyes alone, we are in a position to understand the pattern of life within life—from the largest to the smallest holons.

We know the mitochondria cooperating in our cells long ago worked out a mutually consistent way of life with other cell parts. We know they and we have a mutually consistent arrangement as we provide their fuel and safety while they make our energy. We see how all species, including our own, must work out their mutual consistency with one another as co-evolving parts of the great Gaian body.

○ ○ ○

Of all the cooperative steps in Gaia's dance, the most important was the invention of sex—the sharing of creature plans by uniting DNA from more than one "parent" source in a single new being. We have already seen it at work among monera, where all strains of bacteria have access to a single great DNA pool and sexual exchanges occur constantly without being linked to reproduction, as Lynn Margulis has pointed out in tracing the origins of sex. It is because of this sexual freedom—lost forever in later kingdoms of life—that bacteria are streamlined creatures with an efficient communications system and tremendous flexibility, able to trade information worldwide and thus adapt to any emergency.

In the kingdom of protists, sex took some strange new twists, very likely quite by accident. These twists eventually linked and limited their sex to reproduction and to two partners within the same species. Thus the boundaries of sexual reproduction became our way of defining species boundaries. The within-species sex of the protist kingdom was passed on to multicelled creatures, though sometimes different species co-evolved to help each other in their reproduction, as in the case of flowering plants cross-pollinated by birds, bats, moths, bees, or other insects. But before we get to larger creatures, let's see just how the kind of sex we know—the production of offspring by the mating of two parents—came about among protists.

A prokaryote without a nucleus, remember, reproduces by splitting or budding after its loose DNA loop unzips and copies itself. Even though this process gets a bit more complicated with the great quantity of DNA packed into a eukaryote's nucleus, most eukaryotes, including our own body cells, also divide in this way. The process begins with the unzipping and copying of each chromosome, their neat coiling and recoiling into chromosome pairs buttoned together by centromeres or kinetochores, as we saw in Chapter 6. The nuclear wall dissolves to release them, and the cell constructs a fabric of microtubules on which the pairs of doubled chromosomes line up. These then "unbutton" themselves by dividing their kinetochores, which then button themselves to the tubes such that the two members of each pair of doubled chromosomes ride smoothly in opposite directions, pulled by the kinetochores to opposite ends of the cell before it divides into two with a full set of chromosomes each.

This usual way for cells to divide is called mitosis—*mito,* as in mitochondria, meaning "thread" such as humans invented for weaving. If you could watch cell mitosis through an electron microscope, the neat formation of microtubules called the mitotic spindle and the shuttling of chromosomes along its threads would look like a live weaving process.

Mitosis is a nonsexual, or asexual, way to reproduce. The offspring of mitosis are clones—offspring of a single parent cell usually thought of as being exact copies of that parent. Yet we know that our whole bodies are cloned from the single fertilized egg we began as, and our cells are very varied. The idea that clones are all alike is linked to the idea that sexual reproduction is what brought variety into evolution. But the clones of rare asexually reproducing animal species, such as certain lizards, are quite as varied as the offspring of the more usual sexually reproducing species.

If sexual reproduction did not evolve by natural selection for the advantage of variety, as scientists thought for so long, then why *did* it evolve?

Again we owe the tracing of a story from the ancient microcosmos to Lynn Margulis and her co-workers, who combined clues from earlier biologists and from their own research. Margulis noted that sexual reproduction has three important aspects: the halving of chromosome numbers within each parent, the doubling of chromosome numbers by bringing two parent cells together, and the alternation between these two stages of halved and doubled numbers generation after generation.

How odd to halve chromosomes continually, only to double them again. Margulis investigated this mystery by looking for cases of halving and doubling chromosomes in the microcosm and for ways in which they might have become linked into a single reproduction system. The story that emerged is, like the evolution of eukaryote cells, one that begins with exploitation and ends in cooperative partnership, and once again starvation is the initial motive.

A desperately hungry protist, even today, may resort to cannibalism, and on occasion may fuse the swallowed victim's nucleus with its own. (All nucleated cells will fuse with one another under the right conditions.) Doubled chromosomes may also come about when a protist begins mitotic division and is then unable, for some reason, to finish the process—failing to divide after doubling the number of its chromosomes and fusing them back into a single nucleus. This, too, has been observed.

In either case, the extra chromosomes may work well in times of need but become unwanted extra baggage when things go well. So protists learned long ago—over a billion years ago, which was not so long after they had become protists—to reduce the number of chromosomes again when this was to their advantage. The process of halving a cell's chromosomes is called *meiosis,* which means "lessening." Some protists seem to have become experts at doubling and halving their chromosomes according to the demand of changing conditions from drought to plenty and back.

The fusion of two sets of chromosomes into a single nucleus if they are from different protists—even if one protist eats the other—is a sexual union, but what has that kind of sex to do with reproduction? The halving of chromosomes in meiosis, as we just saw, was a solution to the unnecessary and troublesome burden of doubled or tripled chromosomes. What we call the *haploid,* or "half-set," of chromosomes that seeds, pollen grains, eggs, and sperm contain are half of our normally double, or *diploid,* number—the chromosomes of all our body cells being paired, one of each pair from each of our parents. Far back in evolution this doubling must have occurred as described, and "stuck" as the normal number. When sexual nuclear fusion became linked to the reproductive formation of new generations of individuals, the double number had to be halved before each sexual-fusion-and-reproduction event to avoid doubling the chromosomes mercilessly in each generation, which would have been literally a dead end. Sperm, pollen, and egg cells are all produced by this halving process in such a way that the fusion of egg and sperm or egg and pollen results in the "normal" diploid chromosome sets of animals and plants.

Cannibalism, fusing nuclei and then reducing chromosomes again, accidents of timing, and perhaps other events finally worked themselves out into a smooth system of sexual reproduction—not the most elegant system nature ever devised, but one that has obviously worked well enough in the world of creatures less elegant and less sexually free than bacteria.

To carry out the work of a developing embryo, the DNA of haploid chromosomes from two parents must match stretch for stretch, or gene for gene. Two new species may not yet have branched away from one parent species enough to prevent mating and making babies, yet their chromosomal match may not be good enough anymore for the babies themselves to be fertile. This happens when a horse and a donkey are mated to make a mule, or a lion and a tiger to make a tiglon. Such branching species usually branch by occupying different ecological areas and so do not normally mate, but humans have shown the possibility, though their hybrid offspring are sterile.

The change to the new way of reproducing did not happen all at once. Even today, in fact, some protists—such as paramecia—still reproduce both in the old way of mitotic cloning and in the newer way of sexual reproduction. Paramecia are a good example of nature's experiments with sex and reproduction, having as many as eight different sexes rather than only two, if we so want to label mating or gender types.

Many protists can reproduce with or without sex—that is, sexually or asexually. Sometimes one way serves their needs better, sometimes the other. Diatoms, for example—those lovely tiny creatures with the fancy silica shells—tend to reproduce just by mitotic splitting. They manage this by making their shells in two pieces that can come apart—one piece slightly larger, fitting neatly over the edge of the other to close it, just like a box with a lid.

The trouble is that while all the offspring must complete the half-shell box they inherit, diatoms have learned only how to make the smaller "bottom" of their box shells when given half a shell. That means the split-off diatom that gets the "bottom" has to use it as a top, making a smaller bottom than its parent diatom had. Over generations, then, lines of offspring inheriting bottom shells get smaller and smaller, only those with direct inheritance of the original top shell maintaining the original size. This problem is eventually solved by sex.

Instead of continuing to split by mitosis, they resort to meiosis, producing little packages of single-set chromosomes—*gametes,* or sex cells that have no shells at all and behave the way eggs and sperm do. Thus, when two of these gametes get together, all the DNA plans for a new diatom—including plans for the top and bottom parts of a normal-size diatom shell—are put into action. That way even the smallest diatoms can bring themselves—or at least their offspring—back to full size.

Sexually reproducing protists—giants in a world of bacteria—evolved into a new variety of complex patterns tailored to different lifestyles and environments. The co-evolution of protists with these environments led to one improvisation in Gaia's dance after another, including the protists' formation of a new kind of cooperative that led to multicelled creatures and thus to the remaining three kingdoms of life—fungi, plants, and animals—in all their visibly fantastic variety.

○ ○ ○

We have watched the earth transform itself from a fiery ball to a crusty planet whose skin came ever more alive as giant molecules and enzymes formed, then packaged themselves as bacteria. We've seen living masses of microbes transform themselves as they created an atmosphere and discovered new lifestyles. We saw how the monera collected themselves into the much larger protists, inventing nuclei and stumbling on sexual reproduction. Now our story continues with the evolution of ever larger and more complicated creatures.

Some protists began living together in colonies—sticking together after division rather than floating off on their own, though each protist in the colony was as independent as if it had gone its own way. But just as protists evolved when various monera began working together as cooperatives, the "first builder" protists living together as colonies soon took the next step of inventing a way to communicate with one another.

After all, communication had always played a major role in Gaian life, from the free exchanges of bacterial DNA to the DNA-RNA-protein communication systems developed between nucleus and cytoplasm within eukaryote cells. Now it was time to develop communications *among* eukaryotes, and indeed they found ways of sending chemical messages from one to another.

This chemical communication made it possible for the individual protist cells to harmonize the things they did—say, to beat their cilia tails together in rhythms that moved the whole colony smoothly along in one direction. The ability to communicate soon became useful in many new ways of cooperating, especially in divisions of labor among different cells. Such colonies began the evolution of multicelled creatures.

A most peculiar and fascinating in-between step in the evolution of multicelled creatures from single cells is the appearance of slime molds. Scientists have devoted whole careers to studying these strange creatures that seem to be part protist, part fungus, part animal. Slime molds start out as separate amoebalike creatures, but when their food source runs out, they emit chemical messages that attract them to one another until they gather together into a visible sluglike community. You can see their jellylike mass sometimes on the underside of a rotting log or leaf. Some varieties form a large sheet of jelly.

This sluglike community can actually move itself about like a brainless worm, but eventually it stops and begins sprouting stalks that form fruiting bodies on their tips. These then release spores—tiny dried-up packets of DNA and other cell materials—the way any self-respecting mold would. The spores blow through the air and, after settling in a new moist place, form new amoebalike creatures to start the cycle all over again.

Slime molds thus are capable of specialization and cooperation under hunger conditions, if not otherwise. Note that we have now found hunger as the prod behind the cooperative evolution of nucleated cells, the invention of cooperative sexual reproduction, and the evolution of multicelled-creature cooperatives—all creative responses very different from the competitive struggle with which Darwin tried to account for all evolution.

In other ancient protist colonies that did not lead the double lives of slime molds, some of the cells became specialists at making food, others at catching it, still others at breaking down and digesting food. In some colonies there were specialized cells to move the whole thing about; others contained specialized cells for sticking it tight onto rocks. The first multicelled creatures were cooperative colonies of protists, just as the first protists had been cooperative teams of monera.

In some colonial creatures, certain cells began specializing in the sexual reproduction of the colony as a whole. That meant a few cells could reproduce the whole colony, instead of each cell in the colony reproducing itself. This was a big step in the evolution of colonies into creatures and in the evolution of embryos as a way of starting new generations. For the continued life of the species, all the other cells now depended on the specialized cells that produced gametes—the haploid chromosome packages that diatoms make. We call them gametes because they were not yet either the specialized eggs and pollen of plants or the eggs and sperm of animals. All gametes looked alike, though they had to pair to make a new being.

The development of multicelled creatures, all of which reduce themselves back to single-celled creatures in each generation in the process of reproduction, was also the development of "natural" or inevitable death by aging. Bacterial progenitors do not age and die. Instead, they phase out their physical and genetic identities over generations of offspring. Multicelled plants and animals, however, leave their bodies behind as their genes fade into new generations. Gaia, limited to planetary matter, can evolve only by recycling it. The invention of natural death for larger creatures was an important new way of doing so.

From a microbial point of view, the large multicelled bodies cloned from eggs and seeds have no further value as new generations emerge, though they become excellent sources of food. The fact that death, whether we choose to see it as natural or unnatural, is necessary for life to continue virtually without end has been hard for us humans to grasp and accept. If new creatures kept coming to life without others giving up their lives, the supplies in the earth's crust would soon be used up and the mass of creatures would all die together of crowding and starvation—as we humans are rapidly learning from our successful efforts to delay death.

With the death of creatures so that others can recycle the materials of their bodies, life can go on and on. In fact, the way living things die to make way for new life in Gaia's dance is very much the way things happen in any dance. Every dancer knows that a dance continues one step at a time—that old steps must be abandoned so that the dancer's body will be free to perform new ones, which may then repeat or change the pattern of old steps.

Gaia, our live Earth, has lived for billions of years and has billions more years of life ahead. Our individual bodies will die and be replaced by others, much as the cells of our own bodies are constantly dying and being replaced by new cells. Every seven years or so we become practically a new person through such replacement of cells—yet we never notice this change. In the same way, from Gaia's point of view, there is no death—just endless replacement of old cells in her body with new ones.

Every atom that is now, ever was, or ever will be a part of us will live on somewhere in Gaia's ever evolving dance for billions of earth years yet to come. Even after the earth dies, those atoms will live on as part of our galactic dance, some perhaps finding their way into new living bodies of new planets.

Animals, as part of our own planet, were a marvelous evolutionary development in the face of yet another problem. Some protists, after running out of food, were unable to make their own food from light because they contained no blue-green bacteria—and none of their chloroplast descendants. We are not sure whether our own protist forebears never took in any blue-greens or whether they took some in and later lost them. We do know that plants have always had both chloroplasts and mitochondria, which allow them to make food using sunlight and to burn food using oxygen. Animals, including ourselves, can only burn ready-made food.

This means that plants could—and still do—live their whole lives sitting in one place, making their own food, while animals had to evolve ways of going after their food. Animals, as we will see, evolved all sorts of equipment—from eyes and ears to feet and wings, to heating and cooling systems, to brains for organizing all this complexity—just to help them chase after food, and all because they had no chloroplasts!

Nature, or Gaia, is of course never quite as orderly as we would have her, so she managed to leave around some puzzles such as the giant green clam, a creature that does have chloroplasts and uses them to make emergency energy from sunlight, though it is clearly an animal in every other way.

Among the earliest multicelled animals to evolve from protist colonies were polyps. Luckily, there are still many living polyp species that match ancient fossils and so give us clues to their early evolution. Actually, polyps look more like plants than the animals they are. Sea anemones, which look like flowers, are polyps; forests of coral are huge polyp colonies.

The polyp animal is shaped like a tube with a flowerlike circle of tentacles at one end around its mouth. The other end of the tube is stuck to a rock or to the body of another polyp in its colony of polyp animals. And there it stays. It is a simple animal with a body organized to catch its prey in its tentacles and stuff the food into its mouth.

Still, many polyps have rather amazingly complicated cells along their tentacles. These cells have a special name; we call them *nematocysts*—meaning "thread bags"—because they evolved from ingrown cilia that grew into extremely long, thin, hollow threads and became very specialized in their job. When prey touches one of these surface cells, the long coiled thread shoots out under the pressure of liquid filling it, tangling the victim and paralyzing it with poison barbs. Nematocysts are a wonderful example of the incredible patterns of organization that nature has worked out even within the cells of the smallest and simplest of creatures. And nematocysts are such good self-contained weapons that other creatures, after eating polyps, may not digest the nematocysts but may, instead, keep them for their own use in catching prey.

Polyps reproduce by budding. This job is sometimes assigned to certain members of a polyp colony, which are fed by the others so they can concentrate on their important work. In some species the polyp buds grow up stuck to the parent, but in others something much more interesting happens. The newly budded polyp breaks off, flips over so its tentacles hang down, and floats off into the sea. As it grows, it becomes a glassy bell or umbrella with a softly fringed edge of trailing streamers—a jellyfish, as we call it, though it is not a fish at all. Its proper name is *medusa,* a name taken from the ancient myth of a woman who had snakes on her head instead of hair. Medusae are a much more adventurous stage of polyp life that learned to reproduce sexually. Some species tried having both sexes in the same individual—as flowers and earthworms have them—while other species began making separate males and females.

In any case, all medusae produce female eggs and male sperm, which fuse to make baby medusae. The baby medusa is so different from its parents that it, too, gets its own name. We call it a *planula*. The planula is a long, flattish blob that rows itself about freely for a while using a fringe of cilia. Then it settles onto a rock and sticks itself tight to grow into—a polyp.

The life of a polyp is thus a matter of *metamorphosis*—changing form—from planula to polyp to medusa. Such metamorphosis was later repeated in evolution—in butterflies, for example—like so many other earlier step patterns woven again and again into the later dance. Polyps in countless variety still abound in the seas, looking much as did their ancient forebears. Yet sometime, somewhere in the dim past, some of them became discontented with this three-stage metamorphosis, which always came back to a sedentary phase, and went on to invent more adventurous lives.

9

From Polyps to Possums

One of the many remarkable things about life is its memory. The process of life creates and stores information—not just the kind of information needed to reproduce each new body from a single tiny cell, but remembered information about much of its evolution over aeons of time. The way each of us came into being shows us something of the whole dance of evolution since the time of the first protists. Before we are even born—in just a short nine months—we repeat many steps in a billion years or so of our evolution as earth creatures.

Each of us began as a single cell, much like an ancient protist that began its life by sexual reproduction, as the offspring of two parents. This new-creature cell divided again and again, cloning itself first into two cells, then into four, eight, sixteen, thirty-two, sixty-four cells, and so on until there was a simple ball-like creature very much like a protist colony. This creature in its early stages lived like other protist colonies in a uterine sea—a special bag of salty liquid duplicating the real sea.

Our own embryonic colony of cells soon specialized and turned us into multicelled creatures. If you could watch a movie of your own development—you may have seen films of other embryos developing—you would see the ball-colony change shape as its specialized cells divide again and again. One side of the ball dents in to form a groove where the backbone will grow. A lumpy head appears at one end. Soon it looks like a tiny fish with gill slits, dark eyes bulging in its big pale head, and a tail that starts twitching. A little later it looks so much like the developing embryos of frogs and turtles and chickens and pigs that it would be hard to say what it was going to become! Even when our arms and legs have budded from our bodies, we still look much like other animal embryos.

Slowly we continue our formation as we roll about comfortably in our warm sea. Our tails shrink to nothing, our brains grow bigger, our arms and legs and faces become human. Finally, after nine months, we leave our maternal sea to begin breathing air.

How fascinating that this memory of the Gaian life dance is relived by each of us, reminding us of just who we are, where we came from, how we are related to all other species and to the whole dance of life—the evolutionary dance we traced at the end of the last chapter to our discontented polyp ancestors. Let's continue now to see what made them evolve into more complicated animals.

○ ○ ○

Dull as the polyp's life cycle was at the beginning and the end, the middle stages as medusae and planulae permitted some adventure, some spreading into new environments. Somewhere along the line, certain young planulae seem to have felt their difference and rebelled against settling down and growing into a dull adult life as polyps. They began growing up straight into medusae, simply skipping the whole polyp stage. Others began sexually reproducing themselves even while they were still free-swimming planulae, gradually evolving, generation by generation, into various species of wormlike and then fishlike creatures.

The evolution of new species from the baby stage of a parent species is known by evolutionists as *neoteny*—"stretched youth." Neoteny is a kind of Gaian step backwards when the dance of evolution finds itself at some adult dead end or blind alley, when for some reason the body organization of the grown-up stage evolves to limits that get it stuck.

It seems, for instance, that when a species becomes highly specialized for a particular way of life, it loses variety in its DNA and ends up with a very fixed body structure that can no longer change or evolve. We can see such dead ends in polyps living today, which are still very like their ancient ancestors—having evolved no further. But in some ancient species, as we said, nature took advantage of the fact that baby planulae, which did not yet have the specialized bodies of adults, could still change. We can trace the descent of some free-swimming creatures back to ancestors who were very likely unspecialized planula babies. Later we will see that neoteny produced other interesting new species—including our own!

Among the new parts of the more complex creatures evolving from planulalike ancestors were networks of nerves. These seem to have evolved from in-turned rowing cilia that became tubes for sending messages from cell to cell. Once animals evolved nervous systems they kept improving them and never gave them up, for these communications systems made it possible to organize ever greater numbers of cells within a single creature, just as nuclei had made it possible to organize ever greater numbers of cell parts within a protist.

Organizing now meant actually forming *organs*—grouping cells into body parts specialized for particular jobs, such as guts for digesting food, hearts and blood vessels for circulating supplies, eye spots for telling light from dark and later making images that helped identify food or predators. Remember that animals were pushed to evolve complex organs because they had to hunt for their food rather than make it themselves. The other side of that coin is evolving ways to avoid being eaten yourself.

An extra tube, stiff but bendable, evolved along the main nerves running from one end of early wormlike animals to the other. This protected the delicate nerves against damage. Later the tube wrapped itself right around them to become a backbone. But long before that happened, the tube was useful in another way. Long cells, good at stretching and shrinking, attached themselves to it, pulling the tube into wiggly patterns. That was the beginning of muscles, which gave creatures an important new way of moving themselves—from the inside, rather than by outside rowing hairs or tails.

Clumps of nerves at the head end evolved into simple brains; the eye spots evolved into true eyes with lenses to focus light and retinas to make images of what went on in the environment. Squid and cuttlefish have bodies that remind us of tube-shaped polyps with tentacles, but they have evolved eyes not unlike our own. It is a biological mystery why such wonderful eyes evolved in creatures with so little brain to understand what their eyes can see.

The muscles of squid and cuttlefish evolved into another way of travel—by jet! Their tube bodies take in water and the muscles squeeze it out suddenly, shooting the creature along. They, along with their octopus cousins, also evolved a way of hiding even out in the open, by making supplies of black ink to squirt like a cloud into the sea around them. Octopuses also evolved bigger brains, and are smarter than one would guess.

In some species of early animals, mouths gave up tentacles to grow a single big sucker, as in some eels, or to build jaws and then teeth, which became very popular. Different species experimented with inner bones and outer shells to hold their bodies together and protect them.

So one evolutionary invention led to another, perhaps sometimes letting natural selection improve an old pattern a little at a time, other times by actively reorganizing available genes quite quickly to produce a new pattern. Every time a creature's DNA replicated itself in the process of forming egg or sperm cells, and every time these came together with gametes of the opposite sex—if not at other times as well—there was an opportunity to reorganize the gene information anew.

Modern ways of looking into the past show us that very likely just about all the genes that now exist were developed in very ancient times, just as even earlier creatures had developed almost all the molecules and proteins that now exist. The only new DNA development since early multicelled creatures seems to be the *regulator genes*, which organize simple genes into more complicated genetic patterns. This makes it possible to keep evolving endlessly new patterns out of the same basic genes.

One of the most fascinating of today's biological puzzles is just how an organism's need to reorganize its genes for future generations—say, because of human antibiotic or pesticide attack—is communicated to the regulator genes of gametes in production. We can see that it happens, yet we don't know how. It remains at present a mysterious but real Lamarckian process.

Much about the organization and reorganization of DNA may be better understood when we understand more about the brain. The DNA pool of bacteria, the cellular nucleus, and the brain are all natural systems for receiving, storing, and processing information required by organisms. The efficiency of the Gaian way of life in using the same schemes over and over in ever more complex arrangements suggests that these three systems are likely to have a good deal in common.

We have learned that the brain is more coordinator than

dictator of the body's physiology and behavior—a kind of central clearinghouse and resource center for the body as a whole. Perhaps the organized DNA of cell nuclei is similar, not dictating what the cell does, so much as being used as a resource center by the cell as a whole.

A step at a time, over millions of years—though some steps as we know were faster than others—the seas filled with an incredible variety of living things co-evolving as ecosystems. Or, we might say, from our Gaian viewpoint, the seas with their ever renewed supplies of salts and minerals washed from land rock turned themselves into a living soup of plankton and larger creatures.

Many species continued the "plant" way of life, though they were not yet true plants—evolving seaweed colonies from simple algae. Some had parts that looked like roots, stems, and leaves, but true roots, stems, and leaves evolved later on land. These chloroplast-containing creatures took over bacterial ways of fixing nitrogen, took other building materials from the sea and from the sea bottom, which was rich with decaying bodies, and continued making food using solar energy. Their usable nitrogen and the carbon they took from carbon dioxide and built into their bodies were also used by animals who ate the algae to build themselves.

Some of the carbon was recycled, but much of it was buried in sea floor sediments of dead protists, algae, and animals. Over billions of years of evolution, this process plus a similar carbon burial on land after plants evolved there used up most of the early atmospheric carbon dioxide. It and other early atmosphere gases were gradually replaced by a balance of mixed gases that were almost entirely the products of living creatures and were just right for their continued survival and evolution—gases constantly burning each other up, being replenished, being recycled by living creatures.

Bacteria and protists continued to live among the larger creatures in far vaster numbers, working at the rebuilding and balancing of the atmosphere and the chemistry of the seas, as well as providing the larger creatures with food.

Plants did not become as complex as animals because nothing pushed them—life was too easy. They had no need to go after their food, to see it, to grab it, to digest whole other creatures. Animals, who did have to do these things, tried out many ways of building themselves. Some evolved hard coverings like those of clams, snails, barnacles, spiny starfish, and sea urchins. Some, like crabs and lobsters, successfully tried out legs in pairs. Others, like worms, squid, and octopus, stayed soft. Still other free-swimming forms became sharks and the bony ancestors of modern fish. But no matter how different their shapes, they all evolved muscles to move with, blood to circulate supplies, eyes to see with, and nervous systems with brains to coordinate their ever more complicated bodies.

Together, this great variety of living things created different environments for themselves and for one another on the sea's floor, on its surface, and in the shallows near shore. These environments acted back on the creatures, tailoring species to each other and to the whole environment. So life created its own conditions and these conditions in turn created new forms of life. That is Gaia.

○ ○ ○

The stardust that formed our rocky earth had come a long way in rearranging itself. But while countless small bits of the planet's crust were turning into living creatures, the crust as a whole broke into pieces that slid about on the softer molten insides, like the armor plates of an armadillo. Ever new eruptions of lava pushed the plates apart by adding cooling rock to their edges, while their opposite sides slid under the edges of other plates to make room.

We introduced them in Chapter 2 as tectonic plates, their name coming from the Greek word for "builder." And, indeed, they built the shape of the world we know. The thickest parts of tectonic plates are the land masses we call continents that stick up out of the seas. At the time the dinosaurs first evolved, all the thick parts had moved together into one huge land mass called Pangaea, which means "all Gaia." Ever since then, this land mass has been breaking up to form continents separated by oceans and seas. The Atlantic Ocean is still getting wider year by year as South America and Africa are pushed ever farther apart.

The continents had been separate before they came together as the Pangaea land mass, but they had then dominated what we know as the southern half of the earth. As Pangaea formed and split up again, the continents moved northward and apart from one another. If you could see their movement over billions of years as a film, you would see them riding the slowly swirling soft insides of the earth.

Let's go back to pre-Pangaean times now, to the next big step in the evolving dance of life—the great land adventure, when some creatures took to dry land and continued to evolve there.

Of the organisms that were larger than microbes, the funguses and plants went to dry land first, paving the way for animals by multiplying into rich food supplies—just as bacteria had paved the way for plants by breaking up rock and reproducing themselves, thus starting a supply of soil.

The spores of plants and funguses were blown onto land by wind. Wherever bacteria had made enough soil and there was enough moisture from dew or rain, such spores developed.

The first land plants to evolve were mosses growing close to the ground along with cooperating teams of algae and funguses called lichens, which look like plants. It took a long while for plants to develop strong roots and taller, stiffer bodies that could evolve into ferns and trees.

Perhaps the migration of plants onto land began when some of them were left high and dry for hours every day by moon-pulled tides. To survive, they adjusted their bodies, learning to live both in and out of water. The major new structures they had to develop to support their bodies in air and carry water from roots to all parts were *vascular* systems—stiff tubes, or veins, running the length of plants. The whole plant kingdom is divided into vascular and nonvascular plants.

Animals followed more slowly onto land, when there were enough plants to provide shelter and food. They had to make even greater changes in their bodies to get about on land and breathe the air. Some animals, like the early shore plants, evolved ways of living part-time in the sea and part-time out of the sea. We call them *amphibians*—from *amphi* meaning "double" and *bios* meaning "life"—creatures who lived double lives, on the land and in the sea. In ancient times there were far more species of amphibians than there are today. Most of the ancient species died out as other animals evolved. Among those still living today are frogs and salamanders.

While amphibians have their bony skeletons inside their bodies, *arthropods*—jointed-foot creatures, whose name comes from *arthro* meaning "joint" and *pod* meaning "foot"—have their hard skeletons outside. Some arthropods also came to live on land, evolving into lightweight insects, while their relatives in the buoyant sea evolved the larger bodies of crabs and lobsters.

The first amphibians were fishlike creatures that crawled on fins and developed lungs to breathe air. Eventually they transformed their fins into short legs, and some species evolved into reptiles. Later, some of the reptiles went on to evolve into birds while others transformed themselves into mammals. A few of the mammals later *de*volved their legs back to flippers and fins when they returned to the sea to become seals, sea lions, manatees, whales, and dolphins.

But let's stay for a moment in the great age when reptiles stalked the earth as the largest of its creatures for 160 million years—two to three times as long as mammals have been around, some forty times as long as humans have existed. The fossil record shows many shapes and sizes of reptiles and lets us know what important roles they played in evolution. Looking at dinosaur fossils, we see that today's reptiles, such as lizards and crocodiles, must have descended from the dinosaurs' small survivors. And our birds, we see quite clearly, descended from the still half-reptile archaeopteryx—"ancient wings"—and its cousin the pterodactyl—"feather fingers." (The words "wing" and "feather" both come from the Greek root *ptery*.) Their fossils, as we said earlier, show us leg bones evolving into wing bones, jawbones into beaks, scales into feathers. These early birds were far larger than today's, with wing spans up to twelve meters. People have begun building flying models of them to see how such great beasts could stay in the air.

Dinosaurs themselves were up to twenty-seven meters long and some species towered tall on their hind legs. During the dinosaurs' reign, the world buzzed with flying and crawling insects and grew ever greener with plants, some of which at last developed flowers.

Flowers brought a wonderful new blaze of color into the world, a special kind of beauty, but that beauty was practical as well. With flowers, plants achieved their full two-parent sexuality, but both sexes were housed in the same individual. Flowers, that is, have both sexes in one, but they found a special way to spread their genes around to other potential mates they could not reach themselves. Flowers gave plants a way of attracting birds and insects to cooperate with them in getting the male pollen of one plant to the eggs of another.

The next and last new kinds of animals to evolve were the warm-blooded mammals, who keep their own body temperature constantly warm and keep their babies inside their bodies, rather than in eggs, until they are ready to be born. Late in the age of dinosaurs, some reptiles called therapsids were evolving into mammals, just as the archaeopteryx and pterodactyl were evolving into birds. Though mammals and birds were until recently thought to be the only warm-blooded creatures, we now know that some fish have evolved this feature as well. Red-blooded tuna, for example, keep much warmer than the surrounding sea, with systems that are extremely efficient in preventing heat loss.

The large dinosaurs disappear suddenly—on Gaia's time scale—from the fossil record 60 to 65 million years ago, when some great catastrophe—probably, as we said earlier, a giant meteor or planetoid striking the earth—changed the weather, the climate, the whole environment on which they depended, so much that the huge specialized creatures who had ruled Gaian life for so long could not change enough, or fast enough, to become mutually consistent with it.

Just as many of Gaia's early bacteria had died of oxygen poisoning, all land-dwelling dinosaurs, as well as the largest evolving birds and mammals, were killed off by this more recent disaster. Yet just as new types of bacteria took over in the first instance, so new kinds of animals evolved from the small ones left after this great extinction. The only large creatures to survive were those large mammals that had returned to the sea to continue their evolution as aquatic creatures.

Gaia has shown again and again the ability to recover from disasters, always continuing the dance of life in creative new ways. Or, we might say, rock can rearrange itself in ever new patterns if old ones no longer work well.

○ ○ ○

Life continued, particular species of plants and animals evolving particular bodies and ways of life to balance and harmonize with one another as parts of the great Gaian system in which they evolved—a system that went on working as a single being to regulate the earth's temperature, chemistry, and weather. Living things may even help tectonic plates to move by weighing them down as their bodies turn back to rock, or at least by providing chalky layers that help some of the weighted plates slide under the edges of others.

As Pangaea broke up, splitting Africa from South America and separating Greenland out between America and Europe, the Atlantic Ocean flowed between all these pieces. India, at that time, had "floated" about halfway from Antarctica, where it began, to Asia, where it was to get stuck in place, pushing up the Himalayas as it did.

By the time that dinosaurs disappeared, the earth had begun to look as it does today, though the continents were still closer together than they are now. As land bridges between them disappeared under water, plant and animal species were stranded on separate continents to continue separate paths of evolution. This is why many species alive today, like the kangaroos and koalas of Australia, are found only on a particular continent. Some species of different continents, such as the alligators of North America, the caimans of South America, and the crocodiles of Africa, are still recognizable descendants of the same Pangaean ancestors despite differences due to their later evolution.

Reptiles were, and still are, cold-blooded animals, which means their body temperature rises and falls with the temperature of their environment. When it is cold they slow down; when the sun warms them they become active. That is why dinosaurs hunted their food by day. Some dinosaurs may have evolved into warm-blooded creatures, but the evidence is still unclear.

Like the birds descended from them, reptiles lay eggs. But unlike warm-blooded birds, they rarely take care of their young. Their brains are so simple that most reptiles can't even recognize their own babies when they hatch. In fact, they have been known to eat their own babies—not because they are vicious or cruel but because they cannot tell the babies from other edible things. This challenge to reptile babies made them evolve into creatures that run very fast very early. A baby reptile, in fact, is almost as good as a grown-up at doing most everything reptiles do.

Reptile behavior emerges directly from reptile genes, bodies, and physiology, without benefit of very much thought or choice. They do pretty much the same things in the same old ways, not having enough brain to think about or change them. They hunt food, show off to win their mates, huff and puff at strangers who come into their home territory, and fight if the stranger is not frightened away. That's about all. They don't even sleep, but just settle down quietly when night cools them off.

Some of today's small mammals show links back to the early mammals that evolved from reptiles. The platypus, for instance, is a strange, furry, warm-blooded animal that lays eggs and has a beak and webbed feet, as though it couldn't decide whether to evolve into a bird or a mammal. True mammals branched into two types. One branch, which includes kangaroos and opossums, gives birth to very undeveloped babies that stay in pouches on the outside of the mother's body after they are born, until they are ready to live on their own. The other branch grows its babies to a later stage deep inside the body until they are born.

Possums are one of the most primitive—or antique—species of mammal living today. They seem to have changed little since they evolved among the dinosaurs, so they are important in the study of evolution. Among other things, they may have been the first animals to sleep and to dream.

10

From Possums to People

O We are still not sure why warm-blooded animals sleep and dream. Sleeping and dreaming are special behaviors that evolved and were passed on to all mammal descendants, yet no one really knows why. Scientists used to think sleep restored worn bodies, but there is no good evidence for this theory. Some people sleep very little, even never, and have no problems as a result. For most of us, however, sleep is clearly an unavoidable aspect of our physiologies. That makes it highly likely that sleeping and dreaming evolved because somehow they helped our ancestors survive.

One guess is that sleeping was a way to keep warm-blooded bodies, which worked just as well by night as by day, quiet and in hiding during the hours when dinosaurs hunted. Dream pictures of such dinosaurs may have kept the mammals just close enough to waking so they could get up and run fast if a dinosaur poked its head into the nest. This theory seems to fit mammals that are active by night. But shrews, small mouselike mammals, are active day *and* night, while bats, which are small mouselike mammals with wings, sleep almost the whole day and night. Some mammal species, including humans, do their sleeping at night instead of by day. Another theory says that we are put to sleep by a chemical that daylight builds up in our bodies, making us more and more tired by nightfall.

All this shows that the mystery of how and why sleep and dreaming evolved is still unsolved. All we know for sure is that they were among the many new kinds of behavior organized in the larger, more complicated brains of early mammals such as the possums. Perhaps we sleep simply to prolong our lives, for sleep does slow down body activity and thus makes a body last longer. Bats live about five times as long as their superactive shrew cousins with the same body size. Both species have about the same number of heartbeats per lifetime! But long life in individuals does not necessarily mean their species survives longer, as it is clear that bat and shrew species have survived equally well.

Possums—besides being warm-blooded, hunting by night, and sleeping with dreams by day—take care of their young. Caring for babies was a new and important evolutionary step in mammals. It became possible because the bigger brains had room for more complicated ways of seeing, feeling, and doing—of understanding and acting. Among the new ways of seeing and feeling was the ability of parent animals to recognize their babies and feel the need to care for them—to feed them, clean them, protect them, and teach them, as both birds and mammals do.

Again we must note that nature does not keep things in neat human categories, and so we find insects such as wasps providing for their young in advance of their hatching by storing food, and frogs that keep live young in their stomachs, regurgitating them now and then to see if they are ready to function on their own. Some species of fish have evolved quite elaborate care of live young by their parents, though they are not warm-blooded. Seahorse daddies keep babies in kangaroolike pockets, and some daddy cichlid fishes keep hatched babies in their mouths. After they learn to swim on their own, the fathers herd them and let them back into their mouths any time danger threatens.

Still, if you could watch and compare a fish father and a cat mother caring for their young, you would quickly see that the fish has simpler movements and acts more automatically, while the cat often seems to have a choice about what to do next. The tendency toward particular patterns of behavior evolved in animals together with the structure of their bodies. Such innate, or built-in, behaviors—say, nest building or courting rituals—are popularly called "instinctive," though most scientists, having dropped the old concept of instincts, call them "species-specific" or simply "innate" behaviors. Whatever we call them, the animal performs them without having learned them, and so they appear to be automatic, to require relatively little choice, and to offer little flexibility.

As we move from simple animals to amphibians, to reptiles, and then to mammals, we find nervous systems getting more complex and behavior becoming more flexible. As brains grow larger and more complicated, some of the innate behaviors loosen up, giving the animal more flexibility and choice in responding to its environment and to its own inner urges. The brain is ever more a flexible coordinating system at work in changing behavior to fit changing circumstances.

As animals gained more freedom in what to do, and how and when to do it, they gained more freedom in acting on their feelings and learning new behaviors. Mammals clearly began to show what we recognize as feelings, and some of these feelings seemed to be what we might call the beginning of love in animal evolution. Animal mothers apparently felt good staying close to their babies, feeding them on milk from their own bodies, licking them clean, and hiding them from danger. Whenever mothers and babies got separated, both felt bad and cried. They got back together by listening to each other and tracking the cries. That was the beginning of voice communication, which whales, dolphins, and humans have evolved into language that can be used to say all manner of things. Birdsong is a more innate pattern of voice communication, though baby birds often have to learn the exact pattern for their species from the adults.

One of the most important things mammals learned and taught one another was to play. Play in baby animals seems to be practice for more serious grown-up behavior such as hunting or winning a mate. You can see almost any kind of baby mammal play, from kid goats butting one another to puppies hiding from and pouncing on one another. Mother cats clearly show their kittens how to practice hunting on one another without being too rough, and the kittens come to enjoy the activity. Ever since it evolved, play has been an important part of social life—the life of animals living together, making a living together, communicating, and caring for one another.

Mammals such as cats, dogs, and monkeys have flexible bodies that make it easy for them to care for babies and to tumble about in play. Flexible feet with toes can become very useful paws with which an animal can do many things. Stiff-legged mammals without flexible toes, such as goats, antelope, and horses, can do fewer things, and so they lead simpler lives, grazing rather than hunting other animals, running from their predators, shielding their babies as best they can. While their strong, slender legs and hooves make them surefooted runners, such legs are too specialized for baby care and play, so their young stand and run on their own stiff legs early.

Mammals of all shapes and sizes—some very specialized, some more flexible—branched off from early possumlike ancestors. Each new species created new steps in the Gaian dance of life as it wove itself into a complex environment. While the oceans filled with animals from microbe-sized plankton to sea snails, from fishes to mammalian whales and sea lions, forests grew thick with colorful insects, birds, and furry mammals, plains rang with the clattering hooves of the fleet-footed ones.

With the dinosaurs gone, mammals were the largest of land animals, just as trees were the largest plants. All through evolution, larger and larger individuals had evolved in both the plant world and the animal world. But there was a size limit beyond which individuals did not work so well. Trees that grew too tall could not stand upright in storms or pump water to their highest leaves. Giant dinosaurs were too large to survive a catastrophe that smaller creatures did survive. Mammals the size of whales and elephants seem to be about as big as the Gaian life system can manage successfully. Larger bodies, among other problems, would have trouble getting enough food and oxygen. An animal ten times as big as another in each direction weighs a thousand times as much and needs a thousand times as much food. The large animals that have evolved are very few in number compared with smaller species, down to the most numerous of all, the endlessly hardworking bacteria.

While some Gaian animals reached size limits, there seemed to be no limits to variety in their body designs and behaviors. They evolved endless and wonderful ways of swimming, slithering, running, climbing, and flying; of hunting, fighting, playing, and learning. They had senses to see, hear, smell, feel, and otherwise perceive their world. They communicated with one another and found endless ways of making homes and feeding themselves. They developed marvelous body designs, colors, and dances to attract their mates.

Where could Gaian creation go next? What was possible that had not already been developed?

Think about the world as it was with this tremendous variety of monera, protists, funguses, plants, and animals woven into and weaving the patterns of life. Rock had transformed itself into countless creatures, which had created a rich atmosphere, nourishing seas and soils for themselves. Every species of Gaian creation had its part in the dance as their co-evolution was the evolution of a single great planet body. This great Gaia being knew itself in the same sense that our bodies know themselves, having the body wisdom to take care of itself and to keep on evolving. And yet its creatures lived out their lives, never thinking about those lives or their relationships with one another any more than do our own cells and organs.

Like the cells in our own bodies, these creatures played their parts well without having any idea where they came from or where they might be headed. In a way, they were like actors who had learned their parts in their sleep and played them out with no idea of what the play was about.

And this is just where the next step in the dance comes in—with the evolution of a brain big enough and flexible enough to find ways to communicate with other brains through language far more flexible and complicated than body gestures, grunts, or whistled tunes. Through the social use of language, the new big-brained creatures were to become aware of themselves and one another and of the play they were part of, to wonder about it and about their roles within it.

The biggest-brained mammals in the sea were the whales and dolphins. And what successful creatures they were! Their bodies had evolved in perfect harmony with the sea to which they had long ago returned. They could roam the watery three-quarters of earth. They could "see" by sonar when the water was murky and with their eyes when it was clear. They could adjust to the coldest and warmest, the deepest and shallowest waters, and they could devise a language in which to talk and sing to one another from one side of an ocean to the other.

It is quite possible that cetaceans—whales and dolphins—have evolved the ability to think about and talk about the world they live in. Their language is very complex and apparently very much faster than ours. We have just begun seriously to study it, and so far we have made little progress in understanding it . Until very recently we humans believed ourselves to be the only intelligent creatures, and we didn't treat other creatures with much respect. In fact, we were well on our way to killing off all the whales and dolphins just as, earlier in our history, we killed off most species of elephants—the only animals other than the cetaceans with brains bigger than our own. *Homo sapiens* may even have killed off other species of humans while competing with them for food—the last other known human species having been the Neanderthals, who disappeared from the evolutionary record only about forty thousand years ago. Some scientists think they may have disappeared through interbreeding with our own species; others think we killed them off.

In any case, cetaceans had little to worry about until our own human species evolved brains as big and clever as their own, thirty million or more years after theirs had evolved. Though we humans evolved so much later than cetaceans, by Gaian standards we evolved with incredible speed, our brains blossoming suddenly, expanding in size almost explosively from the much smaller brains of our apelike ancestors.

To become human, those ancestor apes of ours had to do something similar to what their much more ancient polyp ancestors did. They had to practice *neoteny*—that is, they had to remain childlike by not growing up into mature apes at all!

After piecing our own history together from fossils and other evidence, we can see that the story of human evolution goes something like this: Generation after generation, some baby apes were born prematurely; at birth, the bones on the top of their heads were still soft and not grown together. This permitted their brains to keep growing after they were born. In the first year of a human baby's life its brain grows three times bigger and passes the size of a grown-up chimpanzee or gorilla brain. But human faces keep a babyish flatness instead of developing grown-up snouts and jaws and bony eye ridges. If you look at chimpanzee families, you will easily see that people, even grown-up people, look much more like baby chimps than like adult chimps.

We know now that our DNA is 99 percent identical with that of chimpanzees, and that we branched off from our common ancestors only two or three million years ago. Some scientists find it difficult to believe our species can be so different from our large ape relatives with so little difference in genes, but perhaps the major changes in our bodies have been due simply to a few genes that regulate maturation. The potential for greater size and complexity in ape brains may be there, but locked up by the early sealing of their skull bones. In any case, knowing how very closely we are related to our much more peaceable chimp and gorilla cousins may help us overcome the obsessive idea that we are "naturally" violent creatures.

Without snouts in the way, the babyish apes could easily see what they were doing with their paws, which had become very good at grabbing and holding on to things. After all, they had been adapted to a life of swinging through trees for millions of years. Now, more and more, these evolving creatures made their homes on land instead of up in the trees. Their arms and paws, no longer needed for swinging, were free to do new things. In time, their paws evolved into hands with opposing thumbs that were wonderfully useful for holding and making things. Meanwhile, their necks and hipbones, legs and feet, were gradually repatterned for walking and running upright.

The longer these new upright mammals stayed babyish, the later their teeth came out and the less hair they grew, though they may also have lost hair in evolving a cooling system that let sweat out all over their bodies. The cooling system helped them run for such a long time that the animals they learned to chase for food and clothing tired out before they did.

Young humans needed warmth, protection, and affection. They needed to be taken care of much longer than other young mammals. But their long childhood gave them time to play and be curious and learn new things, and parents had plenty of time to teach their children before they grew up. In fact, it is because we never really grow up the way apes do that we can keep on playing and learning new things all our lives—providing we don't seal our own brains up with fixed ideas as inflexible as the old innate behavior patterns.

There were endless new challenges for these first humans. Without the warm coats, strong claws, and long teeth that other animals their size had, they had to survive by using their big brains, their clever hands, and their ability to run longer, if not faster, than many other animals. But perhaps this was in their favor, for even today the most creative and successful people are often those who didn't have everything they wanted early in life and had to work hard to overcome disadvantages.

When roots, seeds, nuts, fruits, and other plant food were no longer adequate, early humans added meat to their diet. And with the challenge of outsmarting other animals by stealing their catches or by hunting them directly, they became more inventive. They made weapons and tools to extend the use of their hands. They also discovered that they could keep warm by wearing animal skins.

We don't know very much else about the lives of the first humans, but we can be quite sure that they were both cooperative and competitive. Families and groups of families shared their caves and their way of life, living together where food was plentiful, traveling together when food got scarce. As they became ever more social and communicative, we can imagine them laughing and dancing with joy when life was good, growing frightened and hostile when food was scarce, when dangerous animals prowled near, or when fires were started by lightning near their homes. Yet there is no reason to assume they were any more hostile to one another than chimps or gorillas are. Even if we became a predator species through hunting, it would have been highly unnatural for us to kill our own kind except under conditions of the most extreme need.

How exciting it must have been when new improvements were invented. Imagine the first people who carried burning sticks to their caves, learned to feed flames without letting them spread and to cover coals so they could be carried to a new home and relit. How proud they must have been to teach such discoveries to their clan or tribe.

Like apes, humans had very flexible, expressive bodies and a natural talent for imitation. Human language probably began with dances to show each other, and later remind each other, of their experiences. We can imagine them dancing hunting adventures, or ways of doing things, or the roles they played in living together cooperatively. We can guess that the sounds they made as they danced took on meaning, eventually becoming spoken word symbols of actions and things, just as drawings became picture symbols. Once there was spoken language, it must have been much easier for them to learn and teach ever more complicated ways of life to one another.

As the ancient bacteria had gotten together to do different jobs within the same cell walls, and as the cells of protist colonies, or the ants of ant colonies, divided their jobs among themselves, so humans now organized themselves into communities where different people did different jobs. Some hunted or fished; others scraped skins for clothing. Some became specialists in making tools or baskets or clay pots, which were hardened in fires. Some were no doubt better than others at drawing, dancing, or telling stories. And some became leaders who organized jobs, drew up rules to live by, and made decisions on what to do when others could not agree.

Human communities, as they got larger and more complicated, evolved leadership and governments just as eukaryote cells had evolved nuclei and just as animal bodies had evolved brains.

Every holon that grows larger and more complicated must evolve some way of organizing itself to simplify and manage its complexity if it is to survive. When human communities got too large, they must have split, or budded off, into new colonies, thus reproducing themselves.

○ ○ ○

We have long assumed that our Stone Age ancestors were brutal and aggressive, using their hunting clubs on one another, gradually civilizing themselves through the invention of better weapons. We have named the ages of man by the metals with which people made their weapons. Because violence plays such a big role in modern human societies, we are fascinated by the idea of early and "natural" human violence. We write books and make movies about it, as though we need to reassure ourselves that violence is the natural way of life for humans. But there is no more evidence, for example, that our species violently eliminated the last other human species, the Neanderthals, than there is that those Neanderthals were peacefully integrated into ours.

Among the oldest artifacts of the earliest Stone Age societies are "mother" images carved in stone or modeled in clay. So many of them have been found by now, without any similar images of hunters or warriors, that it is clear their makers were more interested in the giving of life than in the taking of it.

The giving of life is a lifelong business of feeding people, and early humans gradually learned to bring their food supplies close to hand and make them more reliable by sowing seeds and keeping captive animals. Where climate and soil permitted, they organized themselves into villages with fields. They improved the art of agriculture—not merely planting seeds but selecting seeds from the best crops in each harvest, clearing larger stretches of land. Thus humans began to alter their environment, eliminating some species and nurturing others, changing the genetic identity of plants and animals.

The agricultural revolution of the Stone Age was no doubt the greatest transition in human history, forging our destiny as a species that would change the face of the whole planet, destroying and rearranging things to our desire. New evidence suggests it all began peacefully and modestly with simple agricultural techniques such as keeping herds of animals for milk and meat and wool. In nurturing one another, people learned to spin yarn and weave cloth, to mold clay into larger vessels, to make houses of clay bricks and furniture of wood. They learned to melt metals from the earth to make images of worship, adornments to wear, and stronger tools.

They made musical instruments and danced and sang to celebrate and tell colorful stories about their world and themselves. Some of those stories were reports, others were used as lessons, still more were told just to exaggerate, boast, entertain. We can only imagine the earliest efforts to explain things that were hard to understand—such as what the world of nature in which they lived was all about. From the earliest artifacts, however, we can pull clues to ritual worship of powers beyond humanity.

When we continue the story of human social evolution in the Stone Age and after, we will see that sustained violence among humans was quite likely a distortion of our humanity that came late on the scene of human civilization—and this gives us hope that we may recover from it. But let us first take a look back and reflect on the fact that for the greater part of our existence we were not so different from the rest of the animal world.

○ ○ ○

Looking at ourselves today, it's hard to see humans as natural animals. But from Gaia's point of view that is just what we are—and a very new kind, or species, of animal at that. For most of our history we did little more than shelter and care for our young, gather food, defend ourselves and our homes against other animals, then relax and argue or play with one another before we fell asleep to dream our restless dreams. Compared with the few million years in which human life was not very different from that of other mammals, our civilizations are still a very new development in a still new species.

Imagine squeezing the four and a half billion years of earth's existence into just twenty-four hours. We would find that humans have only been around for the last minute. And only in the last *second* have people formed settled societies. We really have just begun to become human. If we manage to stay around as long as whales already have in their present form, we have many millions of years to go.

Since we evolved, we have survived four ice ages, which were believed until now to have been abnormal times for our planet. Lovelock and others presently working out Gaian physiology suggest that ice ages may be the Gaian norm, with periods between ice ages being abnormally warm, like fevers. We now know of seventeen distinct ice age phases, most of which occurred before humans evolved. Ice ages are only a few degrees of temperature colder on the average than the times between them, though we sensitive creatures consider them great extremes. In any case, during ice ages, great patches of ice move, as we know, from the poles over parts of the earth that are warm between ice ages. As more ice is produced, the sea level is lowered and more land is exposed near the equator—as much land as there is in the whole continent of Africa. When the ice recedes again, these areas are flooded, but ice-covered areas warm up, melt the ice, and produce rich new forests and grasslands.

Humans were apparently driven ahead of the great sheets of ice when they advanced, and humans later followed in their wake as the ice sheets receded. They wandered to places that were rich in food and water and settled down until, many generations later, a new wave of ice drove them back. By the time the last ice age was over, around ten thousand years ago, people had spread themselves out over much of the earth's land, taking with them their early civilizations, their ways of changing the land, their seeds, and their stock. But before we go on with their history, let's look a bit more closely at the spectacular brains that made all this possible.

The Big Brain Experiment

If you could look inside your own head as we looked in on our developing embryos earlier, you would see yet another kind of evolutionary record. The innermost part of a human brain looks much like a reptile brain, and it seems to be this deep core of the brain that sometimes makes us huff and puff and attack others automatically, as though we have disconnected it from the rest of the brain that might give thought to what we are doing. Wrapped around this core, so to speak, is a more recently evolved part of the brain, which, together with the core, looks like a more modern mammal brain such as that of the horse. This mammalian complication of the brain permits those mammalian feelings we call anger and love, sadness and joy. It also appears to make us playful, curious, and eager to learn.

On the surface of the human brain we find the newest complication—the *neocortex*, or "new bark" growth, which ripples and folds itself to look like a great elaborate walnut inside its skull shell. In reality, the human brain is not separable into parts in such a way that we can say that this or that feeling or behavior comes from one or another of its parts. Yet we know it is the evolution of the neocortex, richly interwoven with the inner brain, that permits the whole brain to demonstrate our special human abilities to remember in detail our past, compare it with our present, and on that basis make plans for our future. The neocortex is very much involved when we create our ideas of the world, communicate in language, and think up our inventions, from courts of law to computers, when we become artists and scientists, decide what is good and what is bad, learn and think about our relationship to all other creatures of the earth and even of the whole universe.

There is a great deal that we still don't understand about brains, including their most basic functions, but when we compare the brains of many species we can see quite clearly the ever more complex pattern of their evolution and associate it with the parallel evolution of more complex behavior. Consciousness, communication, and freedom of choice in behavior seem to have evolved gradually in the relatively bigger brained species, yet in our own species and perhaps earlier in cetaceans, there seems to have been an unusual explosion of these talents.

Cetaceans have brains that are bigger and at least as richly complex as our own, and they were that way many millions of years before ours. Like humans, dolphins seem to have ideas, ethics, complex languages, and personal names, though we still know very little about how humans and dolphins can communicate ideas with each other.

If we were to retell the ancient myth of Gaia's dance in light of modern knowledge about evolution, we might say that Gaia danced day and night throughout the ages, creating an endless variety of children, watching over them, and eventually trusting those with the biggest, most flexible brains to learn from their own experience and live by their own choices and decisions. Of those who were freed in this way, no other Gaian species can do as many different things as humans can. No other is so free to choose its behavior at every turn, so clever in inventing and producing technology, so powerful in its ability to kill or protect other species, to destroy or preserve whole ecosystems worldwide. Truly we are an experiment in freedom, and a risky one at that.

Most animals, as we saw earlier, do almost everything they do by innate direction of their behavior. There is little you can teach an ant or a lizard, as virtually all the ways such a creature responds to its environment have been built into it by evolution. The necessary information to develop these built-in responses is stored in the DNA and cytoplasm of the single cell from which the creature was built. The behavior is then carried out by its nervous system and the rest of its physiology in response to its changing environment—the things and events it encounters. In the last chapter we saw that the larger, more complicated brains of mammals freed them from some of their rigid innate behaviors and let them learn new behaviors through feelings and experience. In mammals, innate behavior is still apparent, but feelings and learning and voice communication add variety and richness to behavior.

There are a few things we humans do automatically, as we say "by instinct," such as running from danger or attacking when we feel threatened, seeking mates, feeling love for our babies, and seeking food or a place to lie down when we are hungry or tired. We don't have to think about such things. But we have no innate programs for painting pictures or digging up fossils or building airplanes or hospitals or doing the millions of other things we do.

Our behavior is guided partly by basic needs, a great deal by what our society has taught us, and somewhat by personal choice. What we do by our own choice usually depends on feelings and experience with our past choices, some of which become habit. What we learn from our society depends on how other people's feelings and experience with past choices have been turned into customs or social habits and rules. It is our society, for instance, that sends us to school and tries to keep us out of trouble with others.

Let's consider for a moment this matter of staying out of trouble with other members of our species. In most other social species, built-in behavior patterns keep individuals out of serious trouble with others of their kind. Many species of reptiles, birds, fish, and mammals, for example, make their homes in places or territories they have claimed as their own. When other members of their species come into this home space, the earlier inhabitants warn them with species-specific dances or songs. Usually the intruders back off and leave. Even if it does come to a fight, the intruders know innately to give up before being seriously hurt or killed. You may have noticed that the loser in a dog or cat fight turns up his or her throat to the winner. At this "I give up" signal, the winner's attack stops as if suddenly switched off.

Animals use ritual dances and fights to win and protect their homes and mates as well as to raise their young in safety. The rituals work like a system of rules for living together in reasonable peace—rules of behavior drawn up by evolution as much as was the structure of their bodies. These rules help them balance their lives by spreading each species out over an area of land or sea without crowding, so that all have an adequate chance of getting enough food. In this way, innate territoriality and aggression work for the good of the individual as well as for the good of the whole species.

Just as animals know innately how to share land without killing one another over it, they know innately when they have enough land and food for their needs. An animal may hoard just enough food to get itself through a hard winter, but no animal piles up food or takes land beyond its need.

The price of our freedom to decide our own behavior is the loss of innate rules to limit our own aggression and greed. Like other animals, we have an innate urge to supply ourselves with the necessities of life, to win mates, to make a home for our family, and to protect ourselves and our family against intruders. But, unlike other animals, which know innately just how to act on these urges, we are free to act on them in countless different ways and we must make our own rules for sharing or not sharing the earth's resources with one another and with other species. If we share resources fairly by common agreement, there should be no reason to use aggression. Aggression is not something in us that demands an outlet, as some sociobiologists have suggested; rather, it is a reserve capacity available in situations of real need. Whether such need arises is entirely up to us.

As we big-brained mammals lost the rigidity of innate behavior and gained freedom of choice, we also gained our consciousness—our awareness of what we are doing, our memory of what we have done, our images of what we might do—that is, our awareness of choice. This conscious awareness of past, present, and future makes it possible for us to predict on the basis of past experience what the effects of our behavior will be—and this is our essential guide to choice. It is hard to imagine any evolutionary advantage to consciousness in species that cannot use such awareness to choose how to act. Nor can choice evolve without the awareness required to exercise it. It is likely, then, that conscious awareness emerged and expanded in the course of evolution along with behavioral choice.

The problem in our species seems to be that we have used our consciousness to predict only the *immediate* consequences of our behavior, while failing to consider the broader, long-range results. In other words, we have not yet learned to use our conscious freedom of choice for the good of our whole species over time. Our "civilized" history—at least the most recent five thousand or so years of it—shows that human individuals and whole societies have used opportunity and power again and again to take much more than they need, usually by taking it away from others and killing them if they resist. We lack built-in limits, and we must choose to limit our aggressiveness for the health of our species.

There is now good evidence that early agricultural societies were indeed peaceful, sharing land and other resources without making war. Eventually they were taken over by certain unsettled nomadic peoples who had become aggressive—perhaps for lack of adequate resources. Since that time, there have always been some people who have far more than they need and others who have far less. We simply do not share resources as well as most other species do. But, then, we are still very new, and as we will soon see, there are signs that we may be working this problem out.

Our need to live in societies is as much inherited from our animal ancestors as our territoriality and aggression. But here, too, we are very free to decide just how to act on it—what kind of society to make or build for ourselves.

Insects such as bees, ants, and termites were perhaps the first creatures to evolve highly organized societies. We do not know whether they evolved as societies, similar to cell colonies, or whether they evolved first as individuals and later assembled into societies. In any case, some species build whole cities, make farms to raise plants and other insects for food, are ruled by leaders, have soldiers and workers who tell one another what to do by chemical messages, make wars, and capture slaves.

We are astonished to see the social insects doing so many things that seem so human to us. But actually they could hardly be more unlike humans. Social insects have been living these complicated social lives for millions of years in the same old way, for their hard, external skeletons kept their brains very tiny and unable to evolve. The things they do all their lives, generation after generation, are innately determined. A worker ant cannot change its mind and become a soldier; a queen bee cannot change her mind and run things differently.

To compensate for the smallness of their brains, these insects specialized their functions in such a way that the different specialists together form one social body. Social insects need one another and cannot survive as individuals alone. It's no use trying to keep a single ant as a pet, for it will lose its appetite, get sick, and die. In a way, it is not a whole creature in itself, but an "organ" in, or a part of, its social body—its anthill society. Perhaps an ant is to its society as a mitochondrion is to its cell.

Mammal species, such as our peaceful gorilla cousins, have much less rigid social behaviors than ants, and yet they, too, have innate behavior patterns that preserve the societies they are born into, social structures that have been tested in evolution and have proved healthful for the species. It is interesting to observe that very aggressive species, such as the sacred baboons of North Africa, have a very rigid social dominance system, while peaceful species, such as chimpanzees, show greater equality and opportunity to change roles.

As we will see later, we humans are free to form and test and change our own social structures, and indeed we have tried different kinds of societies in the course of our history. Yet, for all our experiments, hardly any humans today live in a social structure that is truly healthful for all the people living in it as well as for the other species among and around them. But again, we are still very young, and there is every indication that we could solve this problem by understanding our living planet and by looking to our remote but peaceful past and truly desiring a peaceful future.

Our big free brains and our clever hands have permitted us to make dramatic and sudden changes in ourselves and in our world during the past few thousand years—the blink of an eye on the time scale of Gaian evolution, yet long enough for our chosen ways of life and our inventions to have become at once a threat and a promise to our whole planet. We may now have become as dangerous an experiment in life forms as was that other very inventive species, the blue-green bacteria, which learned to make energy from sunlight and which eventually covered the world with poisonous oxygen, killing off countless other species. And we may, after all, prove as much a step to healthy new Gaian developments as were the breather bacteria that found a way to use the poisonous oxygen in the most efficient energy-making process ever invented, the breathing process that permitted us to evolve.

We are not actors learning our parts in our sleep and playing out our lives unaware. We are awake and free to learn what the play in which we are players is all about, and we are free to change the play by our own choice.

We live right now at the most exciting time in our history. It is a time in which we can see ourselves as never before and understand who we are through knowledge of our own history, our evolution, our universe—when we can see what children we still are as a species and what opportunities there are for us to grow up.

We evolved by refusing to grow up as apes, but sooner or

later we will have to grow up as humans. And to grow up as humans we will have to take the responsibility for using our freedom in healthful ways, to help rebalance the great ongoing dance of Gaian creation and to develop harmonious new patterns within it. But to understand just how we might go about this, we need to look at human history not as something separate and different from biological evolution, but as a continuing part of it—as the social evolution of one species within the Gaian life system.

What, then, has been the historical evolution of roving human hunters and settled human planters in the period of known civilization, which now extends back some thirty thousand to forty thousand years?

People in settled agricultural societies, despite generally favorable climates and ways of providing for themselves, surely faced hardships such as floods and droughts or epidemics of disease. Such challenges made humans ever more inventive in their lifestyles, just as similar challenges had made ancient bacteria inventive in theirs. People learned to store food against times of need, to make rules for sharing land and working it, to make medicines from plants, to make canals to bring water from rivers to fields, and to build boats to carry things up and down rivers and coasts for trading with other peoples.

Archaeologists, in studying early human civilizations, are now finding more and more evidence that early agricultural societies everywhere worshiped a Mother Goddess as the giver of life and regarded men and women as equal partners, though the actual management of these ancient economies may have been on the whole the responsibility of women. Such civilizations developed agricultural techniques, other arts, law, and trade over as many as forty thousand years of peaceful evolution—by far the longest part of known human history. We shall return to them in chapters to come.

Four thousand or five thousand years after the last ice age ended, and about the same number of years before Christ, the goddess-worshiping cultures, especially in the Middle East and around the Mediterranean, began to be conquered one by one. Armed tribes of unsettled nomads and hunters on horseback came in waves from less plentiful colder or desert climates, searching for a better living. Ruled by men and worshiping Father Gods, these tribes broke into the more peaceful agricultural settlements, conquered them and, in many instances, stayed to form new, more complex social orders, which eventually grew into warlike kingdoms or empires. Until recently these kingdoms were considered the cradles of civilization, as we did not know of the earlier peaceful civilizations.

Again we are reminded of the primeval Gaian world, where hungry monera forced their way inside other monera to get at their rich supplies, then stayed and multiplied, eventually shifting from competition to cooperation as they formed the much larger and more complicated protists. Perhaps without the invaders, the settled human cultures would have remained like bacteria cloning themselves—producing the same peaceful offspring cultures over and over. Instead, the conquering tribes came in like the invading monera aeons before, taking over their hosts to pursue their competitive interests, staying to build their own empires. If this pattern among humans follows the pattern of the ancient bacteria—and we will see more signs that it does—then we, too, will work out peaceful cooperation to replace our competition with a healthier life for all.

In Gaian evolution the cloning monera took billions of years to change themselves and their environment in ways that permitted the much faster evolution of oxygen-breathing protists and larger sexually reproducing creatures. In human social evolution the goddess-worshiping cultures took tens of thousands of years to develop agricultural ways of changing their environment to support themselves, while the later god-worshiping cultures have changed themselves and their environments tremendously in only five thousand to six thousand years.

Clearly human social evolution sped up and created more varied and complicated patterns from the time of these invasions. Unfortunately, however, the competitive, exploitative situation that went on inside large bacteria before they became nucleated protists is still going on in the human world. The male-ruled conquering tribes took almost all women's social power and status from them, declaring them inferior and setting up other inequalities in society. The records of these invasions are the first clear indication of large-scale violence among humans.

After the invaders conquered these very differently organized societies, they imposed their own social structures and customs upon them, though some of the old ways no doubt persisted in the new hybrid cultures. We humans are creatures of strong habit; our cultural rules, beliefs, and rituals are the glue of our societies. Our very ability to function has always been heavily dependent on our social ideas and structures, and we have therefore fought to preserve them.

As the new larger cultures fought wars with one another, the losers were absorbed into the winners' empires and forced to abide by their customs, though some always persisted and modified the dominant culture. Thus empires grew ever larger. Great empires were formed in this way by Assyrians, Etruscans, Babylonians, Chinese, Egyptians, Persians, Greeks, Romans; Aztecs, Incas, and Mayans in South America; Kush, Nok, and Axum in Africa.

Within each empire, people were kept organized by rulers who laid down laws and kept armies to make sure those laws were obeyed, to enforce order, and to fight wars with other empires. Often such rulers maintained their power by claiming to have been chosen by the gods; some even said they were gods themselves. Most of the ordinary people in these societies were workers and slaves who grew crops and husbanded animals; hauled or channeled water; mined metals and precious stones from the earth; made pottery, tools, and weapons; built cities with huge, beautifully decorated palaces and temples for their rulers; and otherwise transformed the natural environment to human use.

Empire building by male-ruled class-structured societies became the main process and pattern of human social organization, and in one form or another it has continued right up to our present time. In ancient Greece a brief experiment was made in the self-rule of citizens, a form of government that we call democracy—from *demos,* which means "people" or "community," and *kratos,* meaning "government." Later we will look more closely at this experimental male democracy and why it did not last, though it powerfully influenced our whole modern world. For now let us move quickly through history to see its main pattern, to see how we humans used our big brains to continue the task of empire building.

The Roman Empire conquered Greece and later evolved into the Holy Roman Empire, which ruled Europe; the Byzantine, Chinese, and Ottoman empires were formed farther east. Each of these empires fell in its turn, and eventually the human world divided itself into countries as we know them today. But very soon the most powerful of these countries began building their own empires by conquering new territories far from home.

Two human inventions—the compass and the printing press—helped to expand empires across oceans all over the planet and to spread knowledge and culture all over humanity. Soon more machines were invented to make things other than books in large numbers—the age of mechanical industry had begun. The word "manufactured," which meant "made by hand," came quickly to mean "machine-made."

The making of things by machine transformed the whole human way of life, bringing about a new kind of empire and building a new road to riches. The biggest empires at the beginning of the industrial era belonged to the kings and queens of seafaring European countries such as England and Spain. With ships and weapons they conquered peoples in Africa, Asia, Australia, and the Americas, carrying off riches and rich natural resources, making the people slaves wherever possible, and taking over their lands as colonies. The Europeans used these riches to develop their new industrial way of life. Native peoples in Europe's colonies were forced to dig up their earth to mine iron, copper, and other metal ores needed by the conquerors to build machines and transportation systems. They were also forced to give over their land for monoculture food crops and to grow rubber, tobacco, cotton, and wool—all for export to Europe, where other poor, landless people worked at the machines that turned these crops and raw materials into products for sale and who mined the coal that fueled the machines.

In time the colonies began fighting for and winning their independence, thus breaking up the empires. The North American colonies were first to win their independence, and they quickly developed their own machine technology and industry to compete with European empires. The success of the United States, after it won its independence, however, was not typical. In most other colonies, the best land and resources had been taken over by European settlers who had gotten rich by shipping raw materials to the mother countries. After independence was won, these countries continued the colonial way of life, exporting raw materials and food crops to industrial countries and importing machine-made products.

Many peoples who had once cared for themselves independently by hunting for or growing everything they needed to live on their own land are worse off in their now independent countries than they were before the colonial empires were formed. While they are regarded as "backward" by Western standards, they have actually been systematically underdeveloped—prevented from proceeding with their own natural development—to support already wealthy countries. Even after they win their independence the people are forced to continue working the land for others, as they did in colonial times, often farming a monoculture crop or mining a single metal ore or fuel for export. They must buy their food, clothes, and housing with what little money they are paid, so they cannot escape poverty and often suffer hunger and illness.

Such peoples have lost not only their land and natural food supplies but their whole way of life as well—their tribal organization, their nature religions, their arts, and often even their languages. Progress, they were told, meant learning the ways of Europeans and Americans. But though they gave up their old cultures to adopt these new ones, progress for them only meant getting poorer in every way while those who owned their land got ever richer.

Industry changed the way of life in powerful countries as much as in poor ones. The rich were no longer those who owned big portions of undeveloped land, but those who owned machines and bought up land, stripped it, and put it to use in producing raw materials for industry. More and more people were forced to give up subsistence farming and go to work in factories, swelling cities around the factories as the urban way of life became the social standard and the symbol of progress.

Before the Industrial Revolution all humanity lived off agriculture, and few people were rich enough not to have to produce their own food, clothing, and shelter. In today's world, most of us buy almost everything we consume, which has been made, often far away, by others. Even when products are made in our own country, the raw materials in them very likely came from another. The whole world is now tied together by its *economy*—a word coming from *oikos,* "household," and *nomos,* "law" or "management." The human "household," once a local family or tribal unit, now encompasses the whole world. A vast web of transportation and communications lines has been spun around our planet to move about raw materials and finished products and people to manage the global household.

In just a few hundred years, then—much less than the blink of an eye by Gaian standards—our brash young human species with its big brain and clever hands has razed vast natural ecosystems and transformed the environment into a single economic empire covering the whole planet and ruled by the rich industrial countries. Yet both the rulers and the ruled of this empire are politically split into blocs of countries with conflicting, competing schemes for managing this worldwide economy and its people.

○ ○ ○

Our modern world with all its successes and problems seems easier to understand if we look back again and again to the ancient bacterial monera in the process of evolving protists. The exploitation of host bacteria by hungry invading bacteria that needed their resources reminds us not only of the God cultures invading the Goddess cultures but also of our more recent imperialism, in which some countries invaded others and made them colonies. But using up the host, or colony, bacteria's resources killed the host, and so this process could not meet the long-term interest of the exploiters. This is the lesson we are learning after thousands of years of exploiting one another in our struggle to become a mature species—thousands of years, which are as nothing for such a big evolutionary change. The healthy cooperative system that began evolving when invading bacteria produced energy for the host in return for raw materials is paralleled today as developed countries help their former colonies develop industrial energy. The problem is that we have not yet worked out fair exchanges. The powerful countries still demand more political, if not material, concessions from the supplier nations in the Third World than they would demand in a truly cooperative system. But then, the ancient bacteria did not evolve into protists so quickly, either. Only when many bacteria of various kinds had become involved with one another inside the same walls did a cooperative system, including a common nucleus, begin to evolve for the benefit of all.

Human countries have only recently found themselves inseparably linked inside the boundaries of our planet, and we are just beginning to understand what this means. If exploitation and hostile rivalry continue at the expense of cooperation among countries, the new body of humanity may not survive much longer.

Evolution takes time, but when a natural system has pushed itself or been pushed to certain limits, it can reorganize itself with incredible speed. Humanity has now reached such a critical limit. It has also invented everything it needs in order to accomplish a dramatic reorganization into a healthy cooperative body. Particularly interesting is the fact that bacteria invented communications systems prior to organizing themselves into nucleated cells, and that nucleated cells invented intercellular communications systems before organizing themselves into multicelled creatures.

Communications systems, which we humans now have worldwide, are prerequisites to the organization of larger living systems. Transport systems for moving about supplies also play a critical role in the actual organization and function of such larger systems, and here, too, we are well prepared in our transport capability. If the big brain experiment is to be a success—if humanity is to survive as a healthy body, as a holon in the Gaian holarchy—we must use our communications, our information exchange, and our transport as parts of a cooperatively organized body. The sooner we recognize that this is our only viable direction, the sooner we will get on with the task.

12

What the Play Is All About

Earlier we said that humans are the only species in the play of life that can think about and try to understand what the play is all about. Yet we are just now forming a scientific worldview in which we understand our world *holistically*—as a whole made of interconnected parts. We are just beginning to understand how we are related to all the other players in our planetary holarchy of holons; to understand that we are new players in Gaian creation—new players with the responsibility of exercising the freedom of choice our big brain gives us in ways that will keep the play going for all of us.

This idea of humans as part of one huge cosmic play, with freedom we must learn to use responsibly, is actually not new, but ancient. The Hindu Vedists and Chinese Taoists understood things this way, as did Homer and the first Greek poet to write plays surviving to the present—Aeschylus, who lived in the fifth and sixth centuries B.C. The plays of Aeschylus are all about the role of humans in their social and natural world—all about the human task of making responsible choices within the situations and limits set by the world of human society within the larger natural cosmos.

In fact, this playwright's layered cosmos can easily be seen as a natural order of holons in a holarchy. Aeschylus understood how each human choice in behavior affects not only the doer but the doer's family, the society, and the larger world beyond humans.

He saw that the extent of our free choice within the natural cosmos to which we belong is the most remarkable thing about us, and his plays are about the questions people must weigh in making their choices, the effort to understand the consequences their choices will have at all cosmic levels.

The ability to think about choice—to make images of our relationship to our world and imagine the consequences of alternative choices we can make at each step through our world—is the biggest role of our conscious mind, which, as we said, evolved in our animal ancestors along with behavioral freedom of choice. The ways in which we picture our world and our relationship to it are our worldviews, and these naturally have a lot to do with the kinds of choices we make in the play of our lives.

To understand what human worldviews are and where they come from, let's begin by considering how other species view their world, for whether it has eyes or not, every creature has some way of "seeing" its world—some way of getting information about its environment. Without such information it could not interact with that environment and therefore could not function at all.

We have said that every creature is a holon within the larger holons on which it depends. To live, it must "know," in some sense, what supplies to take in from its environment and what wastes to return to its environment. It must do what it can to protect itself from harm and to do whatever else will help it go on living. Even a microbe can tell whether it is in a plentiful environment or not, can tell what is harmful in its environment from what is not, can tell what is useful in its environment from what is not, and so on. Whether it coordinates all this information to help itself survive we really have no idea, but if it does, that pattern of information would be its simple but useful worldview.

The point is that some kind of environmental information or knowledge is necessary to the survival of any living creature, even if its "choices" are not freely made but programmed into it by evolution. In fact, as different creatures evolved, different worldviews evolved. The worldview of a microbe is clearly not the same as that of a marsh grass or a mongoose. Every living being has a worldview tailored to its own needs and experience. This is because each creature is a system capable of interacting with its environment through its unique ability to take in information and act on it.

No creature, even with a brain as sophisticated as ours, "sees" what is really out there in its world. Our eyes do not "photograph" the world. There is nothing remotely like a photographic mechanism in our eye-brain system. Inside our eyeballs, light does strike our retina in a way that reminds us of a photographic plate or film, and the light pattern does produce a related pattern of nerve signals that travel to the brain. These nerve signals, however, are soon joined by a far greater number of other signals coming from inside the brain itself, combining the brain's own information with the incoming information to produce our visual images. What we actually "see" is this complex production of our brains.

The same thing happens when we look at a photograph. The reason the photograph resembles what we saw with our eyes when we took the picture is that our eye-brain system responds in the same way to the scene as it does to the photo that is a mechanical copy of its pattern of light.

Let's "look" at how a frog sees its world. A frog lives mainly on insects it catches as they fly or swim or crawl by. Its eyes and brain and body are an automatic system that has evolved to see and catch bugs. Whenever a tiny dark speck moves across the piece of world the frog's eyes are aimed at, it shoots out its long sticky tongue and pulls in the tiny dark thing. This system works very well in the frog's natural environment, because tiny dark things moving about that way are almost always insects.

We can fool a frog into trying to eat tiny shadows that we move past it, or into actually eating buckshot pellets that we roll past it. The frog does not learn that the buckshot is inedible, but will keep going after it until he is too heavy to move. Its catch-moving-dark-specks system is built into its brain and cannot be changed by experience. Its worldview is not subject to change in a way that permits the frog to learn which dark specks are edible and which are not.

A human baby also puts all sorts of things into its mouth, but for the baby this is a way of finding out what is edible and what is not, how things taste and feel. Its tongue, eyes, ears, and other sense organs form a system that determines and limits the kinds of information it can receive, but the baby is not programmed to put the information together in very fixed ways. The baby must test its world constantly, learning about it through direct experience of it, as well as through what it is taught and told. So, in time, it *builds* a worldview. From infancy on, our brain tests each new experience against those we had before in order to keep the worldview we are building consistent.

Let's look at another example of species differences in worldviews. Suppose a child is playing with a cat when it sees a bee land on a pretty flower. And suppose that all three—the child, the cat, and the bee—can talk to one another.

"What a pretty pink flower you have chosen," the child might say to the bee.

"What pretty pink flower?" the bee might well ask. "Can't you see I have chosen this flower for its deep red stripes? This kind has my favorite pollen."

"Red stripes?" the child says. "I should think a bee could see better than that! This flower is pink as pink can be."

"I beg your pardon," says the bee, "but it is *you* who do not seem to see very well! Look here, cat, is this flower striped or am I crazy?"

"Both of you are nuts," says the cat, "making a fuss about a flower. They all look alike to me, and rather dull-looking gray things they are, sitting about as they do. Now a grasshopper is another matter . . ."

Such an argument might occur because each of these three actually does see the flower differently—the child sees it as pink, the bee perceives it as red striped, the cat sees it as dull gray. Bees can see more colors than humans, while cats see scarcely any color in their world. Bees need to know the world of flowers in order to make their living, but flowers do not matter a bit to cats.

Bats can "see" in the dark, and dolphins in murky waters, by bouncing sound waves off objects in their environment. Dolphins and dogs create a good part of their worldviews from sounds we humans cannot hear; most mammals live with a worldview made more of smells than of sights. Birds and insects sense the patterns of magnetic fields in the atmosphere.

Each species has its own system of senses that bring patterns of stimulation from the environment to its brain as it explores and does things to its world. These patterns coming in from outside merge with the existing inside patterns to create the perceived image or worldview our minds are capable of projecting "out there," thus fooling us into thinking we see things the way they really are.

The interesting thing about the projected worldviews different species see, or otherwise sense, is that none has a truer picture of the world than any other. A worldview made of smell patterns is no less true than one of light and sound patterns. Each species has evolved a way of constructing its worldview that best helps it get along in its world. Each gets only part of all the information, or patterns, available in our world—the part it most needs to survive in health and do its part in the greater dance of life.

Only we humans know that all these different worldviews exist. We can record and measure light waves beyond those we see, sound waves beyond those we hear, chemical smells beyond those we smell, magnetism we do not feel. We alone can understand that our perception of the world involves only a small part of all the information available and do what we can to expand our range of information. We have figured out how to peek in on the worldviews of other species by using instruments that copy the way they perceive or sense their world. With sonar, we can "see" as bats and dolphins do; with microscopes we can look in on the world of microbes; and so on.

Surely this, too, makes us a special brain experiment—the only species of players able to understand something of what all the others are doing in the play. We humans have, in fact, a strange ability that no other species has, as far as we know—we are able to make ourselves the audience of the very play in which we are acting.

No other species is in a position to find out what all the others are like and what they are up to. No other knows itself as observers of the whole world, indeed the whole universe. Nor did we ourselves when we first became human. In fact, it is likely that we played human roles in the play of Gaian creation long before we could stand apart from the world in our own minds to see ourselves and others as players. What was it, then, that changed our worldview to the perspective of audience?

In bacteria and protists, there is a pretty direct link between stimulation from the environment and behavior in response. Their sensory parts are directly connected to the parts that behave by contracting, rowing away with cilia, or engulfing a food particle. In multicelled organisms, the stimulus may occur many cells away from the moving, behaving parts, so communications systems evolved—chemical and hydraulic systems in protists, funguses, and plants, ever more elaborate nervous systems in animals. But the more complex the nervous system became, the more it developed its own patterns to come between the incoming sensory patterns and the outgoing behavioral patterns. The connections, that is, are no longer direct, and the creatures' worldviews are determined more by their nervous system's species and individual history than by the new patterns that actually come in from outside.

In social species something else comes into play between the senses and behavior—the whole history of interactions among socially related individuals. There is, in a sense, a social brain or mind organized and shaped by social interactions and language over time, incorporated into the brain and behavior of individuals as they learn to live in society.

Language has played an enormously important role in the building of human societies and cultures. The human mind itself is largely a product of our social language community. Language is without doubt at the heart of our humanity. And written language may have been the invention that changed our mental images of ourselves and our world more than any other.

It was very likely writing that changed the way we saw ourselves in relation to the world in which we live—that permitted us to be observers of as well as participants in life's play.

Before the development of writing, it did not occur to people to think of themselves as separate from their world—as observers, or knowers, of it—the way we do today, when we can "see" the history of the world in a book, or another part of the world in a film, which is the latest variety of "book."

Before writing, language was not a "thing" in itself. Talking was simply a skill like walking. Nor could people imagine knowledge being passed on through language in any way other than through direct learning from another living person. Neither poetry nor law nor any other body of knowledge could exist without a live human knower of it before words could be carved in rock, inscribed on a clay tablet, or written on papyrus. Writing made us observers of our own play and gave us a way to store information and pass it on unaltered to our own and all future generations.

It's hard for us to imagine what it must have been like not to have separated ourselves from the world as observers of it, not to think of ourselves as separate from our knowledge, not to think of our languages as languages and our minds as minds, or our world as something to know *about* in our minds. Yet all of us were like this as small children before we were taught to read and to think about the world. In this sense, human infancy even today repeats something of the infancy of our species.

Before we invented writing, the script of Gaian creation was stored in geological records, in DNA, in the development of embryos, in nervous systems, to some extent in human minds constructed by language. Through writing we began to separate our knowledge and ideas about the dance from the dance itself—in a sense, to separate the script from its playing out.

But remember that the "reality" we humans see as ourselves

and our relationship to the world around us is our creation, our conscious imagery. Our worldviews are rich in the images that language makes possible—an ever increasing wealth of linguistic portrayals of our human interactions with one another and with the rest of nature, accumulated in time and over cultures through written records.

Even our pictorial art is influenced by the way the artist's mind "talks" about what it sees. Try imagining something without imagining accompanying labels and ideas. The flower we saw earlier through a child's eyes as quite simply pink and pretty has become, as we say, all things to all people: a tasty morsel for a farmboy to feed to his donkey, a symbol of beauty and a token of love to a lover, a source of perfume to a maker of cosmetics, a model of reproduction to a biology teacher, a solar energy plant to a physicist, a warning of fading youth to one who is advancing in age. And any one of us can change these *pre*ceptions into other *per*ceptions as often as we like.

The meaning we give things comes from the context in which we see them, and we supply this context not from the sense impressions we receive from our world but from the patterns ever evolving inside our nervous system—patterns which reveal themselves as that richly complex self-organizing and partly conscious mind which ever composes and recomposes itself through individual and cultural experience.

It is our human heritage to continually work at making conscious, thoughtful sense of all these patterns. We embed all new information into existing brain-mind patterns, put these patterns into categories or contexts of meaning, add to them, change them, rearrange, distort, and enrich them until they make sense to us as part of our overall worldview.

We have wanted and needed to make sense of the world for as long as we have known anything about ourselves as humans, and we can do this only by using our free minds to create meaningful worldviews. Yet just because our brain-minds are so free, each individual human can see and understand the world as differently as do various other whole species. If all of us saw and understood the world in the same way, without being told anything about it by others, we wouldn't have to try to *make* sense of it, or try to teach one another just what kind of sense we think it makes. Everyone would just *know* how things were and what they meant. We would be like frogs, all of whom see dark specks in just the same way and know just what they mean and what to do about them.

Human worldviews must be created through the personal experience of living in the world. At the same time, all of us must fit our personal experience into a worldview given to us by others. For if our own experience does not fit into the way others see the world, we will be thought quite mad. In fact, people whose worldview is completely different from the one we call "normal" are commonly considered insane.

Only by agreeing with one another on what the world is all about—on how to make sense of it—can we have human societies or cultures. Most of our worldview actually comes from our culture—from family, friends, schools, books, television, and so on—though all of us add our own special touches through personal experience and ideas.

Our creation of worldviews is thus yet another way in which our brains are an experiment in freedom. While most other species have evolved their special way of seeing the world as they have evolved their behavior—building it into their body plans—we are free to improvise both our worldview and our behavior. When we look at human history to see what a people's worldview was in a different time and a different place, we see that worldviews have evolved along with the visible aspects of culture, and that there is a very powerful relationship between the worldviews that people hold and the kind of society they construct—an inseparable relationship, that is, between the way people believe their world *is* and the things they *do* to one another and their world. In practice, our worldview is our script for the play of life, assigning each of us our role within it.

The most important kinds of worldviews we humans have created are religious and scientific worldviews. In religious worldviews, a goddess or god—or both or many deities—creates the world and then continues to rule or look out for it in some consciously purposeful way. In purely scientific worldviews, both ancient and modern, the world creates and runs itself without conscious purpose.

In both kinds of worldview, we humans see our job as understanding how the world is ordered—by what god-given or natural laws it works, or what else gives it a meaningful or at least coherent pattern. The way to this understanding, however, is very different in the two worldviews. In religious worldviews, the order is learned from certain people, such as priestesses or priests to whom it has been revealed, or from sacred writings such people have recorded. In scientific worldviews, the order or pattern of the world is learned from scientists, who look for it in nature, make theories about it, and do experiments to test their theories.

Theories are no more, no less, than well thought out ideas or models of what various aspects of the world seem to be and how they work. Scientific theories are ordered worldviews that can be tested against predictions we make from them, though we must expect them to change as we ourselves change and as we gain new knowledge. Until recently, religions had worldviews that were not to be questioned, but with new historical information some religious worldviews are now changing to come into greater harmony with scientific worldviews.

Religion and science thus give us our worldview—our image of the natural world and our relationship to it—but our worldview also includes our ideas of what our relations to one another are, or should be, so it includes political and other cultural images.

Until this century, it did not occur to people that they could have anything to do with creating their worldview. All through history, people thought the way they saw the world was the way the world really was—in other words, they saw *their* worldview as the *true* worldview and all others as mistaken and therefore false. Disagreements between people who believed in a particular religious or political worldview and people who didn't believe in it—who had a different worldview that suggested different kinds of behavior and social structures—have been an important cause of many wars—for example, the Christian Crusades against Muslims in the Middle Ages, the democratic revolutions against kings in the past few centuries, and the more recent communist revolutions.

People are very reluctant to change their worldview, because their worldview holds everything together—as we said, it makes sense of the world. To change a worldview is to lose that sense, and so a worldview usually changes only by force—when people with one worldview conquer those with another, as in the ancient conquests of Goddess-worshiping societies by God-worshiping tribes—or by persuasion, as when missionaries or scientists persuade people to adopt a new worldview in place of their old one.

Perhaps the most important discovery of modern science is that *there can be no single true and complete worldview*. Like all species, we have only partial information about the world, and our information changes as our knowledge increases, as our inventions become more sophisticated, and as we and other species actually change our world. We change the world even while we are looking at it, for we are never only observers—we continue being players.

Most of what scientists do is try out—test by experiment—different parts of a scientific worldview, to see where it works and where it needs changing. Archaeology and history have become scientific searches now, along with biology and physics, seeking experimental ways of testing theory. It makes good sense to keep improving our worldview as we gain new knowledge, even though it will never be completely "right."

There is no scientific test for the truth of our worldview, but there is scientific and practical evidence showing how well we get along in the world when we use a particular worldview as a map to guide our experience and behavior. We can test our worldview to see whether it guides us to as healthy an existence for ourselves and the larger life system of which we are part as do the "worldviews" that are programmed into other species. We will say more on this in later chapters. For now let us note that for the first time in our history, we as individuals are consciously evolving our worldview by thinking about, questioning, and testing it rather than just letting the events of history and a few powerful individuals dictate it to us.

○ ○ ○

Part of our evolving scientific worldview, as we said, is recognizing the validity and importance of other species' worldviews, expanding our own by incorporating theirs. It is equally important for us to recognize the validity and importance of different *human* worldviews, expanding our own in so doing. Every human culture that has its own language and customs has ways of seeing the world that are unique, though any human individual can learn any human culture and language. We can thus communicate across languages and share cultural experience in ways that enrich us all.

Today's dominant scientific worldview evolved in European languages with common roots and close relationships—languages which happen to be structured in a way that forces us, in talking or writing about our world, to think and speak of it in terms of thing-nouns and action-verbs. This language structure gives us a worldview, as soon as we begin speaking as children, in which we actually *see* the world as made of separate things that stay still (nouns) or move or are moved in relation to one another (verbs). The reasoning, the logic, and the mathematics of scientists are all based on this way of dividing up the world.

It can come as a great surprise to us that all people do not see the world in this way. Some human languages do not make our kind of distinction between nouns and verbs—all the world is seen through certain languages as a pattern of interwoven processes in time more than as a pattern of separate things in space. Speakers of Hopi or Nootka, for example, cannot imagine things without their motion, change, aging, or other aspects of coming into and out of being. Instead of saying, for example, "The light shines," or "The water falls," they have single-word expressions that do not separate the light from its shining or the water from its falling. In such process languages, people do not think of time as made up of a series of "things" called seconds, minutes, and hours. They see time as the change in things, which is the way physicists now understand time.

These are only a very few examples of the way in which a language can determine how we see our world, yet they are enough to make us think about what the scientific worldview might be like if it had been developed in a very different set of languages. Einstein once agreed it might be easier to describe the discoveries of modern physics in the Hopi language than in English, because we would not face the contradictions of a world made at once of particle-things and wave-actions, of matter-things and energy-actions, never having separated thing from action in the first place.

Process-language cultures are better suited to organic than mechanical worldviews. Perhaps such cultures did not develop mechanical technologies because machinery is necessarily conceived and built as the interactions of separate and, insofar as possible, unchanging parts. As it happened, science developed most fully in European-language cultures along with machinery, becoming closely associated with it, as we will see in greater detail shortly.

Anthropology and linguistics, the sciences of human cultures and languages, are relatively new parts of science as a whole, but they have taught us that human experience is very varied and rich. They have made us realize that the scientific worldview, which was developed mainly in industrializing Western countries, is based on and represents only a limited part of human experience. Many scientists, especially physicists and physicians, have begun to use ideas and practices from Eastern worldviews to enrich their Western science, while Western science has become an important part of Eastern life, especially in Japan and China. Unfortunately, just as we dominant humans have worked at killing off other intelligent species, so have our dominant technological cultures worked at killing off nontechnological human cultures. Every culture and language lost seriously diminishes the variety and richness of human experience. This variety and richness is as essential to our cultural health as is genetic variety to the health of any species and species variety to the health of any natural ecosystem. Our human mania for reducing variety to create monocultures is not expressed only in our agriculture and animal husbandry, but in our own cultures as well—and it is equally destructive in all cases.

Keeping in mind this perspective on our still-evolving scientific and cultural worldview—our continuing effort to understand what the play of life is all about—let's now look at its historical roots, its evolution within the cultural traditions of our species. Let's look, in other words, at the scripts people have written for themselves as they played out the historical steps leading us to our present conception of the play.

13

Worldviews from the Pleistocene to Plato

The earliest worldviews we know anything about date back to the Pleistocene epoch of ice ages—to what we call Paleolithic, or Old Stone Age, people—and it seems, as we said, that these people commonly saw nature as a great mother, a goddess who gave them life and all that was needed to sustain them. Seeing nature as the great Mother Goddess would have made sense of human experience by explaining nature's gifts of food and the birth of creatures, bright sunny days, like the goddess's good moods, and angry storms or droughts, like her bad moods. Nature seemed to love and to rage at her human children, giving them reason to celebrate her life-giving and nurturing, to love, fear, and respect her.

We can safely assume that Paleolithic peoples took all nature to be as alive as they themselves were, and that they felt themselves to be part of, or the children of, this great being. Even today, peoples who live in natural settings without changing them significantly tend not to divide nature into living and nonliving parts. Their only concept of "nonliving" is of something dead that has been alive.

Stone Age and Bronze Age peoples must have called the Mother Goddess by many names in many places—only the later names, surviving into written language, were recorded. Just a few that were recorded are Nammu, Utu, Inanna, Ishtar, Iahu, Astarte, Kali, Isis, and Matrona. Iahu, meaning "exalted dove," was the goddess's Sumerian name, and was later turned into the masculine Jehovah. In ancient Greece they called her Gaia, as well as Eurynome and Demeter and Pandora, meaning "giver of all gifts."

The Mother Goddess of these ancient religions was surely not conceived of as outside nature; she was not the external creatrix of nature but the creative force of nature itself. Nature was felt as the creative-destructive dance of life and death. Various components or forces of nature—sun and moon, winds and seas, those animals that were important to people—were assigned to members of the goddess's "holy family" or seen in other supportive roles.

The image of nature as a providing mother and the worship of this Great Goddess very likely influenced the development of Stone Age societies as agricultural "households" whose members provided for themselves and one another by raising crops and keeping tamed animals. Archaeologists James Mellaart and Marija Gimbutas, as well as the archaeological scholars Merlin Stone and Riane Eisler, have given us an entirely new image of early civilizations, as exemplified by the well-preserved Neolithic town of Catal Hüyük in Turkey.

Most striking in such well-planned and -managed agricultural societies, with their large towns, agricultural technology, wall paintings, decorated pottery, sculpture, and other arts, is that unlike later cultures they show no evidence of fortification, warfare, conquest, slavery, or any significant social inequality, judging by house size, burial customs, and so forth. This is taken to mean that men and women worked in partnership, and there is evidence at Catal Hüyük that those in need were provided for from public stores of food or from the goddess's temple gardens.

Such ancient societies seem to have practiced the kind of peaceful democracy our modern societies are still far from bringing about. The remains of cultures throughout the Middle East, North Africa, and Europe, including pre-Minoan and Minoan Crete, show advanced democratic, peaceful, egalitarian societies, in which, as historian Riane Eisler puts it, "linking, not ranking" predominated.

The extent to which these societies were designed and managed by women will probably be debated among archaeologists for some time, but there are indications that women's roles were at least as important as men's. Mellaart found the holy family at Catal Hüyük represented in order of importance as mother, daughter, son, and father, and a similar order was suggested in households by the sleeping platforms, that of the woman being more fixed and prominent than that of the man. There is no intrinsic reason to doubt that women, as the human representatives of the goddess, were accorded the social status that men gained later as human representatives of the god, but these early societies show no indication that men were oppressed by women; on the contrary, they indicate an equal partnership.

If women did have the authority to make social rules, we might expect those rules to have been based on partnership for the simple reason that women give birth to and raise both girl and boy babies without considering the one better than the other, if the mothers are permitted to act on their natural feelings. The creation stories of ancient societies often told of man and woman having been created together, as they were in the original Hebrew-Christian Genesis before it was rewritten to have Eve created from Adam's rib.

On the contrary, the hunter or nomadic Father God worshipers who invaded and conquered these societies were apparently neither peaceful nor egalitarian, but headed by men who were experienced in the use of weapons. Perhaps they had been driven to violent competition by their harsh environment and had come to worship lightning-bolt-wielding and thundering sky-gods because they were so much in the open, relatively unsheltered, vulnerable to marauding attacks by other, similar tribes.

When these conquerors stayed to rule a settled society where they found life gods, they changed not only the social structure and rule but the society's worldview as well. Often they turned the Mother Goddess into the wife or daughter of their chief god and joined lesser gods and goddesses from both religions into a single *pantheon*—or "all god" religion.

Sometimes they got rid of the goddess altogether by making up stories in which the god was great and the goddess was only a disobedient mortal woman who was forever making trouble. Pandora was so demoted. Her name still means "giver of all gifts," but in the story we hear about her she brings only troubles into the world by disobeying the Father God. Similarly, the Hebrews, whose difficult wandering existence in the desert had led them to believe in a stern Father God, turned the Mother Goddess, along with her symbols—the serpent of wisdom and the tree of knowledge—into Eve, another mortal woman who brought trouble into the world by questioning male authority, by disobeying God.

Later, when Christianity replaced the "pagan" god religions, old male deities were also contemptuously dethroned. The Celtic sun god Lugh, for example, became Lucifer, angel of light, who was cast from heaven in medieval times to become Lucifer, or Satan, the symbol of evil.

All in all, the historical record tells us that when some men acquired the power that accrues to those who have weapons and wealth, they formed a worldview based on a belief in their own superiority. They projected their self-image into an authoritarian and violent male god, thus justifying the domination of women, who came to be seen as the property of males, to be safeguarded and bartered. Nowhere is this more graphically recorded than in the Hebrew-Christian Bible, and even the Golden Age is tarnished by a similar treatment of women.

It followed that ruling men extended the idea of superiority and the practice of violence into their affairs with one another as well—making war, dethroning both the gods and the goddesses of the conquered peoples, making warriors their heroes, taking slaves, building class-structured societies.

Another important aspect of the shift from a worldview based on partnership to one based on domination, as god-worship replaced goddess-worship all over the civilized world, was the idea that nature was separate from both gods and people—that it had been created and was ruled over by a god who was external to nature. Nature, as this god's creation, was then seen as a gift given to his people to use and exploit for their own ends.

And so, at the same time—a few thousand years before the Christian era—humanity seems to have undergone the two greatest changes in history since the advent of agriculture. One was the shift from the worldview and culture of partnership to that of domination—from the worship of life-giving to the worship of life-taking, as Eisler puts it. The other change was the shift from a worldview in which people and their deities were part of nature's own improvised dance, continually self-creating from within, to a worldview in which men and their gods stood outside and above nature, in which men claimed the god-given right to exploit women and all the rest of their god-given natural world.

All this, of course, is a sweeping simplification of history for the sake of seeing broad patterns. The role of women as the "glue" that holds society together cannot be doubted even today, but the partnership status, the equal valuation of their work and their arts, has never been regained to this day.

One of the latest goddess cultures to survive was that of Crete, known to the ancient Egyptians as the Keftiu, to us as

Minoan. A peaceful agricultural people, the Minoans left us exquisite art in admiration and praise of nature. Nature goddess worship was evident in other parts of Greece, too, and lasted in some form until classical times—even Plato being initiated into the Eleusinian mysteries of Demeter. But the sequence of Greek myth, as Robert Graves pointed out, shows the gradual destruction of equality and goddess-worship in favor of patriarchal rule and god-worship.

Only the earliest known Greek philosophers were deeply influenced by ancient ideas of nature's self-creation, as were early philosophers in other lands. By the sixth century before Christ, most of the civilized world had been organized into large kingdoms or empires with generally patriarchal religious worldviews and strict laws for keeping order. Yet thinkers such as Lao-tse and Confucius in China, Vedist Hindus and Buddha in India, Zoroaster in Persia, and Thales, Anaximander, and Heraclitus in Milesian Greece (now Turkey) all came to very much the same idea about how nature works. In observing and thinking about nature, they all saw it as alive and forever changing from within, whether or not it was symbolized by a pantheon of gods and goddesses. They saw nature as striving to create its own balance and order through an endless dance of opposing forces or principles such as male and female, light and dark, hot and cold, inward and outward, storm and calm, creation and destruction. In this dance, opposites clashed or simply got out of balance so that things grew, say, too cold or too stormy or otherwise disorderly. Yet somehow new forms and patterns created themselves to bring about new balance and harmony.

Even though they could not talk to one another, these great thinkers all over the civilized world of the sixth century B.C. agreed that nature's constant movement was away from disorder and toward balanced order. This balance, or harmony, they believed, must ever be re-created from imbalance or discord, very much as it is in human affairs. This did not surprise them, because they all saw humans as part of nature.

In Greece such thinkers formed the first scientific worldviews by trying to understand and explain the world solely in terms of what they could see in nature. Poets, meanwhile, continued to spread the religious worldview of Gaian creation and the gods and goddesses who ruled the world, as Homer did in the *Iliad* and *Odyssey*, which had, by then, been written down.

The new scientific thinkers came to be called *physicists*—as the Greek word for "nature" was *physis*—or *philosophers*—from the Greek words *philos*, meaning "lover" or "friend," and *sophia*, which means "wisdom." The wisdom they loved and sought after was an understanding of how nature works, because they believed that by understanding the natural order they would come to understand how to order human life, both personal and social, more wisely.

In the myth of Gaia, the goddess comes out of chaos to transform her body into the earth by her dance. Originally, *chaos* meant "nothingness"; later it came to mean anything that seemed to have no pattern, that was completely lawless or disorderly. The opposite of chaos was order, which was called *cosmos*, the Greek word for "world." The world, in other words, is the pattern of things.

The Milesian philosophers agreed that the natural world is orderly—that it has a pattern that can be discovered, described, and understood by human beings. As scientists, they looked closely at the natural world. In the patterns of stars and planets, the cycles of seasons, the beautiful forms of plants and animals, they saw orderly rhythms, balance, and harmony. This order was disturbed by disorder, or chaos, but it seemed that wherever disorder arose, order was quickly restored. Birds and worms ate up dead animals; old leaves disappeared into rich new soil; rain made droopy plants grow healthy and flower; new forests grew from burned ones. Nature kept making orderly patterns out of chaotic disorder. And what was so interesting about all this was that everything in nature played its part without being told what to do. This observation came to play a very important role in Greek politics before long.

Plants took form, growing from seeds, then rotted back into soil, losing their form. Older animals died as young were born in an endless chain of life. One creature ate another to live itself. Nature was one great intertwined pattern in which, as the philosopher Anaximander said, "Everything taking form incurs a debt which must be paid by dissolving again so that other things may form." He saw this as a kind of justice—each thing, or creature, in nature borrowing from nature's supplies, then paying them back. Rivers dried up while new rivers formed elsewhere. Clouds formed, dissolved in rain, and left clear skies, which later formed new clouds. Fires and storms created chaos, yet from the chaos of destruction, new life and new order always arose. Everything that took on its own form later gave way to other newer forms.

Anaximander's teacher, Thales, thought all things in nature had formed from water, and had water as their essence. Anaximander himself, seeing the fossils of sea creatures on land, thought about the great changes in geology and in life forms that must have happened over time. He came to believe that living creatures first formed in the seas, later came out onto dry land and shed their shells. Humans, he reasoned, must have been born from earlier animals, since the first human babies could not have taken care of themselves. As far as we know, he was the first scientist to see a pattern of evolution by actually observing nature.

The way in which nature was understood by Anaximander, his teacher Thales, and his pupil Heraclitus—all of whom were Milesian Greeks—was very much the way scientists are beginning to understand it again now—as a great dance of life in which all natural things are connected and constantly improvise their steps as they move toward balance and harmony.

These philosophic ideas remind us of the time when people actually lived in democratic balance and harmony within the larger context of nature. Some memory of these times must have persisted, for the poet Hesiod, around the same time as Homer, wrote (as quoted in Eisler's *The Chalice and the Blade*) of a former goddess culture he called the Golden Race: "All good things were theirs. The fruitful earth poured forth her fruits unbidden in boundless plenty. In peaceful ease they kept their lands with good abundance, rich in flocks and dear to the immortals." This race, Hesiod continues, was later conquered by a lesser race of silver and then by "a race of bronze, dreadful and mighty, sprung from shafts of ash," bringing war. "The all lamented sinful works of Ares [God of war] were their chief care."

It is interesting to note that in the original myth of Gaian creation, the Olympic gods were born of the giant Titans, who were the first children of Gaia and Ouranos, the sky whom she created. Were the Titans perhaps a symbolic memory of the powerful patriarchal tribes—the Achaeans who destroyed the goddess's rule at Delphi and put Apollo in her place, the Aeolians and Ionians who overran Greece in later waves?

In the sixth century B.C., the ruler of Athens, Solon, put into practice philosophical ideas of natural balance and perhaps also the mythologic-historic memories of greater equality. Trying to create some semblance of democracy, he divided land more equally among people and made laws to ensure greater justice and to give citizens more say in the decisions of their society.

The playwright Aeschylus, whom we mentioned earlier, also used these ideas in his dramas, giving his heroes and heroines the task of balancing the scales of justice in working out their personal lives within the larger framework of family, society, and all nature. But we also see in Aeschylus how this process of justice was undermined in the shift from a goddess culture to a god culture by the devaluing of women. At the beginning of Aeschylus' famous trilogy about the Mycenian house of Atreus, Queen Clytemnestra's murder of her husband Agamemnon is personally justified by *his* previous sacrifice of their daughter, Iphigenia, and socially justified by the queen's status as head of her clan, with the responsibility for avenging bloodshed. At the end of the trilogy, her son Orestes is tried by the new court of Athens for murdering his mother in revenge and acquitted on the grounds that he has not shed kindred blood. Athena, no longer the ancient nature goddess, but the warrior child of Zeus, sprung to life from his ear, presides over the trial and casts the deciding vote. As Apollo explains, "The mother is no parent of that which is called her child," but "only nurse of the new planted seed that grows."

This was the beginning of the Golden Age of Greece, which all the world remembers for its beautifully harmonious temples and theaters, for its sculptures and Olympic Games, and for its experiment in male political democracy.

It is important to understand that when this limited democracy was formed, the Greeks had no concept of perfection in their worldview. Their traditional gods and goddesses were seen, like people themselves, as part of nature—imperfect, moody, and mischievous, often intentionally creating disorder, from which they then made order under the higher law of justice. Nor did the Milesian philosophers, the first scientists, see nature as perfect. They knew well that nature was never perfectly balanced or harmonious, but always struggling *toward* balance and harmony. Wherever it was won, it was soon followed by new imbalance that drove the dance forward in search of harmony.

If nature reached perfection, its evolution would come to a stop. If things fell back into complete chaos, creation would also cease. Nature's balance is always somewhere between the two—away from chaos and toward harmonious balance. There was certainly no reason in this Greek worldview to expect men or their society to be perfect. On the other hand, there was hope that neither men nor the society they were trying to balance would fall into complete disorder.

The experiment of the Greeks in trying to make order out of chaos by ruling themselves democratically instead of letting rulers tell them what to do was not an easy one. Would men be able to agree on how to balance their society and live harmoniously? Balance in society would mean allowing all citizens to take equal part and responsibility in how things were run. Harmony would mean a love of the good life not only for oneself but for everyone else, too. Men would have to make choices that were good for themselves and, at the same time, good for all society.

While the Athenians struggled toward democracy for imperfect men in an imperfect world, the Eleatic philosophers on the other side of Greece from the Milesians were forming a new kind of worldview. They had become fascinated with the human mind itself, and with a kind of perfect order it had created, a perfect order they had found not in nature but in the man-made language of mathematics.

The Greeks had used mathematics to measure the size and distance of things, to map the heavens and the earth, to make calendars, and to predict events such as eclipses, as had other ancient peoples such as the Egyptians and Babylonians. These were practical uses of mathematics. But mathematics was a wonderful invention by itself because of its perfect order.

Arithmetic and geometry were languages of number and form built on rules of balance and harmony—but not the kind of natural harmony that falls into disorder and has to rebalance itself by creating new forms. Mathematical rules *kept* things in perfectly balanced order, and mathematical figures—circles, squares, pyramids, and other geometric forms—were the most perfect things that humans had ever created. Furthermore, their perfection never changed, never fell into disorder.

Nowadays we know there are no such perfect forms in nature, but some ancient philosophers, including Pythagoras and the Eleatic philosophers who lived in what is now southern Italy and Sicily, thought there must be. They decided the Milesians must be wrong about a natural cosmos evolving through the ever changing, or *dynamic,* balance between chaos and order. The Eleatics developed a very different worldview, of a cosmos created in the mathematical perfection of unchanging balance and harmony.

They decided that nature only appeared to be imperfect because people were somehow blind to its perfection.

In this worldview, the stars and planets were held on perfect invisible spheres that turned around the earth, which was at their center—if they were not fires shining through holes in such spheres. Pythagoras' discovery that musical harmony obeys mathematical laws gave birth to the idea that the heavenly spheres created music in their turning.

In this worldview, there was no evolutionary change—only a perfect turning around and around, a perfect repetition of the same cycles in the heavens, and of the same cycles of birth and death among the creatures of earth. Everything had been just as it was now from the very beginning of the cosmos. In place of the uncountable opposites unbalancing and rebalancing themselves in the Milesian worldview, Empedocles proposed four unchanging elements that mixed in different proportions to form things. Parmenides even arrived at the idea that nothing in the cosmos or world moved at all—that the whole appearance of motion in the world was an illusion, a strange trick of human perception. And Zeno, whose mathematical puzzles still fascinate mathematicians today, "proved" that the world had neither motion nor parts.

Parmenides and Zeno formed these ideas by using their minds to think about how things *must* be beyond their appearance. There cannot be anything that does not exist, Parmenides reasoned, so all the cosmos must be filled with things that *do* exist. And if it is completely filled with existing things, there can be nowhere for them to move. His pupil Zeno was even better at showing logically that nothing is as it seems to be. In using pure thought to decide what the world was all about, these philosophers abandoned scientific observation of nature.

Other philosophers, even if they did not agree that there was no motion at all, did come to feel strongly that the human mind could understand nature better just by thinking about it than by observing it through their own senses. Democritus, for example, came to the idea that everything in nature is made of invisible tiny hard bits he called *atoms*—from *atoma* meaning "individual"—and that the motion of atoms combined them into the things we see. These atoms were eternal and perfect, as they never changed and could not be destroyed.

More and more the philosophers felt that senses such as sight, hearing, and touch, through which we get our everyday experience of the world, were less trustworthy than the reasoning mind. The reasoning mind, which had invented arithmetic and geometry—rules for ordering numbers and shapes—now invented logic—rules for ordering human thought, for making it as much like the perfection of mathematics as possible. Thoughts and ideas ordered by logic could be written down, compared with other philosophies, polished, and perfected. Such a philosophy could exist, like arithmetic or geometry, as a thing to be known in itself and passed on to future generations.

Of all the Greek philosophers, the one who was most fascinated by the way the human mind formed its ideas was the Athenian Socrates, who acknowledged the priestess Diotema as his teacher, as Pythagoras had acknowledged Themistoclea. Socrates was much less interested in what the natural world was and how it worked than in what the human mind was and how it could be made to work better than it did in most people. If ordinary people, not only philosophers, could get clear on just what they meant by the good life and good government, for instance, they might figure out how to improve their lives and their government.

The democratic government in Athens had become a kind of shouting match in which the cleverest speaker won the day, whether he really knew what he was talking about or not. Some men were looking out only for themselves and seeking ways to gain power. If they argued loud and well, others too lazy to think for themselves would vote their way. Many citizens balked at having to listen to one another's harangues and vote on them. Squabbles continued and men complained about their responsibilities as citizens; others took advantage of the confusion to try to bring back dictators. The playwright Aristophanes wrote some very funny plays about solving social problems in spite of the lying, cheating, lazy, yet clever Athenians. He also resorted, in several of his plays, to the idea that women might solve the social problems of government and war better than men could.

Socrates, meanwhile, spent all his time cornering people and pressing them with questions to help them think more clearly, and he was much loved and admired for this by his followers. But his criticism of the muddle democratic government had gotten into finally led to his trial and execution.

Socrates saw people as throwing their ideas together from anything at all that came into their heads—like builders trying to make a building without a plan, throwing together any materials they happened to find lying around. So he tried to teach them to decide just what they wanted to think about, how to recognize and throw out useless ideas, how to make muddled ones clear, and how to build the clear and useful ones into the understanding they sought—in other words, how to think in an orderly way, or how to reason *logically*.

His pupil Plato was influenced by Eleatic philosophy and fascinated by the beautifully clear ideas or definitions Socrates was able to arrive at by reasoning. Plato felt that such perfect ideas must really exist somewhere behind the muddled world we normally see. Not only was the material world beyond the senses perfect, with ideal forms of chairs and trees and everything else in our sensory world, but so also was the world of ideas such as justice, love, truth, and beauty. Our senses shackle us, said Plato, showing us merely the flickering shadows of a perfect world beyond—a world we can reach only with our minds.

Plato's ideal world fit in with other philosophers' ideas of the cosmos as a construction of perfect spheres, a world built from perfect atoms. All the older worldviews seemed childish in comparison with this elegant new one "discovered" by reasoning minds. What could a perfect world, or cosmos, have to do with an unpredictable, moody goddess or with disorderly gods and goddesses who lied and tricked one another to get out of the messes they made? Poets began making fun of the old religion while philosophers began thinking of a new one.

A perfect world had to be the work of one perfect creator, Plato reasoned, a god who existed apart from this shadow world, who created a perfect and unchanging world, *a god who was always doing geometry!*

The old philosophy of nature as alive and creative in its imperfection was replaced by belief in the perfect and rather mechanical creation of a single, though yet unknown, god. Perhaps this new worldview was comforting to the Athenians at a time when they were having so much trouble working out democratic order. At least they could believe in a perfect world just beyond the mess they were stuck in.

14

Worldviews from Plato to the Present

Eventually a new religion came to Greece from the East, a religion that fit Plato's worldview very well. Not only did it explain the creation of the world by a perfect God; it also explained the disorderly ways of humans as disobedience to God, their Father. This gave people a new hope—that they could make themselves and their society more perfect by obeying God's law.

According to this Hebrew-Christian worldview, brought to Greece by Jesus' apostle Paul, God had created the world only a few thousand years before, with all its different kinds of plants and animals, just as it is now and right at the center of the universe. This view was very much in keeping with Plato's brilliant student Aristotle's static view of nature, Anaximander never having been taken seriously on the subject of evolution.

The scenario is familiar: God created the world as a paradise for humans, setting the first two people into its perfection. They were expelled after Eve disobeyed God's law by tempting Adam to join her in the sin of eating the fruit of the tree of knowledge, thus bringing disorder and strife into the world forever after. Still, paradise could be regained after death, in a heavenly world, if people became perfect in God's eyes again, and this perfection could be accomplished by seeking forgiveness and obeying God's law.

In the ever expanding thrust of empires, Rome conquered Greece. Though Jesus himself had preached equality for all, including women, being opposed to any kind of dominance, the holy texts were rewritten to suit the priesthood of the new church after the first Christians had been tortured and killed. By the time the Roman and Byzantine empires split, the church's revised Christian worldview, adapted to a dominance society, had been officially adopted by both empires, with slight differences.

The rich heritage of ancient scientific discoveries—that the earth moved around the sun, that nature was alive and evolving, that humans were descended from simpler creatures, that many of their ills were curable by medicines and surgery—was destroyed, forgotten, or denied as the new worldview took over. In ancient times, when the library at Alexandria was repeatedly sacked and burned by Romans, Christians, and Muslims, nearly a million book-scrolls of human knowledge and culture were lost.

In Europe, for well over a thousand years, all ideas—scientific and other—that did not reflect the Christian worldview were outlawed. Crusades were led against Muslims, and women were burned to death as witches all over Europe for practicing the older nature religions and healing with natural medicines. Christian priests, who explained God's law to people and enforced it, were the rulers of all Europe. Even kings bowed down to the highest priest of all, the pope of the Holy Roman Empire.

Many of Plato's ideas about education and politics in a perfect society were put into practice during the Christian era. Plato had written, for example, that a perfect society should be ruled by the most educated of citizens, people from all walks of life who lived simply, without personal possessions. Though Plato had not advocated the exclusion of women from education and rule, the Christian monks and priests, who wielded so much influence in Europe, otherwise met Plato's requirements. The highest-ranking clerics did indulge themselves, however, in affluent comforts and fine robes. The idea of heaven and hell as places where people would go after death to be rewarded or punished for their earthly behavior also came from Plato's writings. And the works of Aristotle taught Christian Europe formal logic and the pursuit of virtue—which originally meant excellence and later came to mean obedience.

For more than a thousand years the Christian Europeans educated only boys and men, teaching them the invented mechanical languages of mathematics and logic in the dead language of Latin. A dead language is one no longer in common use, no longer imbibed by children at their mother's knee. This point is important to remember when, in the next chapter, we consider how the differences between natural "live" languages and artificial nonliving languages affect our worldview.

For now, let us note that the mechanical languages of mathematics and logic played a central role in the "rebirth" of science in Europe. This rebirth was of course part of the larger rebirth—or Renaissance—of human curiosity and culture that began about five hundred years ago. Through trade with the East, some Europeans, such as the Medici family in Italy, became very wealthy and very worldly. To have beautiful and interesting things around them, they hired architects, artists, and scientists to create splendid new works and to seek new knowledge.

Traders and refugees brought copies of surviving ancient scientific writings from Constantinople, from Arab lands, from Moorish Spain, from wherever they had been preserved, studied, and further developed, as they were especially in Muslim mathematics, astronomy, alchemy, and medicine. These manuscripts reawakened interest in questions of planetary movement through the skies and of the location and nature of the earth, with all its living plants and animals.

Giordano Bruno, the first philosopher-scientist to revive the ancient notion that the earth moves around the sun, was burned at the stake in the year 1600 by the Christian priests of the Holy Inquisition. But only ten years later Galileo Galilei built the first telescope and with its help showed that the earth does revolve around the sun and so cannot be at the center of the universe. Neither Bruno nor Galileo ever meant to disprove the religious worldview, but only to improve it. Yet Galileo, too, was punished by the church. Narrowly escaping the stake, he was imprisoned and forbidden to teach.

It is important to remember that all the founding fathers of modern science were religious men eager to show the glory of God by giving people a better understanding of his wonderful creations. They imagined God very much as Plato had—as a geometer. Mathematics, Galileo said, was the language in which all nature was written. And so the most important task of reborn science was to discover the mathematical laws by which God had created the world.

○ ○ ○

Throughout the Middle Ages, the Renaissance, and even later during the Age of Enlightenment, as Carolyn Merchant documented in *The Death of Nature*, a belief in nature as alive, personal, and mysterious, persisted especially in and around the tradition of alchemy. Yet modern science in its evolution weeded out these ideas in favor of a belief in nature as an impersonal mechanism that had to be brought under human dominance by rational understanding and mathematical description.

The ancient Greeks, especially Archimedes, had begun to make mechanical models of how things in nature work. In fact, the words "mechanism" and "machine" come from the ancient Greek words for these models. Although Archimedes built actual—very ingenious and successful—machines for fighting wars, he felt so ashamed of these crude imitations of nature that he never wrote about them himself. Because the later Greek philosophers had this idea that real machines were poor imitations of the geometer God's works, ancient Greece did not develop much technology.

The Renaissance, however, gave rebirth not only to art and science but to mechanical engineering as well. Scientists themselves began building mechanical models to help work out the geometric patterns of the heavens wheeling around the earth, later of the earth and other planets wheeling around the sun while moons wheeled around planets, all against the greater background of wheeling stars. Each heavenly body they knew was attached, in these models, to its own ring of metal in the great sphere of the universe. Some of the most elaborate models with the greatest number of rings looked a bit like huge sculptured balls of metallic yarn. All these rings wheeling about within one another suggested that the universe must be something like other Renaissance machinery—something like the clockworks in church towers, with various wheels turning other wheels in their mechanisms.

René Descartes—another founding father of modern science—invented a new mathematics along with a whole new framework for the religious-scientific worldview. In this view, God was not only a mathematician, but a Grand Engineer. Using mathematical laws, God had not only created a cosmic clockworks but had put into it endless smaller mechanical inventions such as plants and animals and people. Descartes insisted there was no real difference between man-made machines such as clocks, grain mills, or jeweled golden wind-up birds that sang and the living mechanisms God had created. This soon became the dominant worldview of all science.

As men were God's favorite mechanisms, Descartes explained, He attached to them inventive minds that worked quite like His own. Women, like animals, had no such minds, and were to be controlled by men along with the rest of mechanical nature. Now, since men's minds had been made to work like God's own, it was no surprise that men, too, were inventive engineers, putting their own mechanical robots into their own mechanical clocks high upon the church towers in honor of the Grand Engineer.

None of the scientists who accepted this worldview ever seemed to wonder if man had not put his own mind, talents, and achievements into his image of God rather than the other way around. They were convinced that, except for their own male minds, everything in nature was God's mechanical creation, made to be understood by men. Surely this was as strange a worldview as humans had ever held, yet it wove the religious worldview and the scientific worldview into one and gave scientists new visions of understanding and controlling all nature, as they were convinced God intended.

Imagine the excitement they felt—if everything in the whole world, even in the whole universe, was mechanical, then men who understood mechanics could understand how all nature worked by taking everything apart to see what made it tick. And sooner or later, surely, they would be able to make their mechanical birds as good as God's.

Another founding father, Francis Bacon, who is credited with having developed the scientific method, wrote much about the coming Golden Age of Science, when man would understand and control all nature, creating his own mechanically perfect societies, which would be free of all human problems. Wasn't that what God intended for his favorite creatures? Bacon, who was a lawyer and who attended many witch trials, saw nature as a woman. Science, he said, would flourish when men grew up and stopped expecting her to unveil herself at their request, but instead hounded nature and tortured her secrets from her.

It is interesting to contrast Bacon's vision of a future Golden Age with ancient Hesiod's lament for the times of a past Golden Race—to reflect on what seems utopian in the context of different worldviews. For Hesiod peace and bounty seemed to pour from nature itself, whereas for Bacon all good things were to be wrested from nature at any cost by dominating and controlling it.

○ ○ ○

The mechanical worldview suited the next few centuries very well, for nothing mattered more to the Europeans and their offspring Americans than the machinery that was changing the whole human way of life, as we saw earlier. Scientists, living in a society that was becoming ever more mechanized, saw more and more mechanisms wherever they looked in nature. Geologists described geological mechanisms—how the earth was put together, how the cycles of weather ground up rocks, and so on. Biologists spoke of the mechanisms of living things—how the parts of plants and animals and people were put together and how they worked. In time, doctors spoke of heart and lung pumps, of bone and muscle mechanisms, while psychologists started talking about the machinery of the brain and social planners worked on the mechanisms of society.

Scientific discoveries of natural mechanisms depended on the invention of new man-made mechanisms such as telescopes, compasses, thermometers, barometers, scales, clocks, and later more sophisticated devices, all of which made it possible for the scientists to detect and measure ever more parts of the natural world. Science also depended on the invention of new mathematics for modeling relationships among these measured parts of the world, for only measurable parts of the world could be studied by scientists with a mechanical worldview.

Let's look for just a moment at this role played by mathematics in science. Mathematics itself is not a science; it is the art of making complicated and beautifully balanced patterns from very simple basic symbols and rules for combining them. Mathematicians can keep finding new patterns to make from the basic symbols and rules they have adopted, or they can change those symbols and rules to develop a whole different set of patterns.

In pure mathematics, the symbols have no real-world meaning—no one has ever found a 2 or a + or a > in nature. But scientists have found that when they assign a real-world meaning to the symbols, some mathematical patterns turn out to be very useful as models of the measurable parts of nature they study. Mathematical models are actually much like machine models, except that they are easier to make from symbols than from hardware.

Consider that whenever we want to describe something previously unknown to us—when we want to understand it, that is—we must find something familiar to which we can compare it. The brain can use new incoming patterns of information only in the context of its existing patterns. Every day we say things such as "This material is as soft as a baby's skin and as blue as the sea" or "That man is watching me like a hawk" or "The boss wants everything to run like clockwork." But how often do we reflect on the fact that all new understanding, even scientific understanding, comes about in this way? The patterns or forms of machines and mathematics are human inventions, and thus they are very familiar to their inventors. So when scientists use them as models—saying, for example, that "hearts 'pump' blood" and "plants 'pump' water" and "$E=mc^2$"—they are telling us how to understand something about hearts and plants and energy.

The idea that reality is made up of only measurable things, the description of which is the only possible knowledge of reality, is called *positivism*. The tasks of positivist science have been seen as twofold—to discover the parts of natural mechanisms and to see how the mechanisms work through the movement of these parts. In other words, scientists took things apart in order to see how they were constructed as well as what made them tick. This method of reducing things to their parts came to be known as the *reductionist method* of science.

In so reducing things to their parts, scientists showed us a fascinating inner world of things. Live bodies, for instance, were made of bones, skin, blood vessels, nerves, and other organs; each organ was made of tissues, each tissue of cells, each cell of chemicals. In fact, everything turned out to be made of chemicals, which in turn were made of molecules.

Molecules that were fixed in tight patterns so they could not move formed solid things, such as rock or wood. If they slipped and slid around each other, they formed liquids. If they floated about loosely, they formed gases. And molecules, it turned out, were made of atoms. At last scientists had the instruments and the mathematics to show that all natural mechanisms were really made of those tiniest and most indestructible building blocks called atoms—just as the ancient Greeks had said more than two thousand years before!

Except that things did not prove to be quite so simple. Atoms turned out to be made of parts themselves, and their parts turned out to be anything but stable and solid machine parts. Atomic physicists, just when they reached the very foundation of mechanical nature, discovered that nature is not so mechanical after all. But before we go on with this story, let's go back to look at the other part of what scientists were doing—seeing what the movement of parts was all about in natural mechanisms.

From their measurements and models scientists worked out mathematical laws of motion among the parts of natural mechanisms. And the more they studied motion in the universe, the more the universe seemed to move and change. Not only did the earth no longer stand still at the center of perfect heavenly spheres, but it turned on its own wobbling axis and wheeled around its sun, which in turn wandered through space as part of a galaxy.

Geologists, digging into rock and studying landscapes, discovered that the earth itself has changed a great deal over time. Biologists meanwhile grew curious about the fossils geologists uncovered. Earth seemed to contain her own record of plants and animals that had lived long ago, and the record indicated that they had changed a great deal, too.

How could the earth be only a few thousand years old, as measured from the generations of people listed in the Bible from the creation of Adam to historically known kings? The geological record was proving it to be very much older, with different kinds of plants and animals at different times in its history. Had God created these different kinds of plants and animals at different times, rather than creating them all at once? Did he keep making them more complicated with each wave of creation? Or had they somehow changed by themselves?

Once the idea of evolution, buried since Anaximander's time, emerged again, it quickly made a great deal of sense, and the whole scientific-religious worldview was turned upside down to fit it. Creation had been seen as a kind of ladder with God at the top. On the next rung down were the angels, then people, then the large animals, then the smaller ones, and on down to lowly worms and even smaller things—all earth creatures having been created at once by God. In the new evolutionary view, the ladder began at the bottom with the most ancient, tiniest creatures, which changed over time, forming new rungs of ever larger creatures, climbing the ladder up to the rung of humans, who seemed to be at the very top—for scientists were beginning to doubt the existence of angels and God.

In Darwin's theory, remember, natural selection worked through competition for limited resources, so that only the fittest survived. As the industrialists of Darwin's England were in just such competition for survival with one another, they adopted the new evolutionary theory as part of their worldview. The rich industrialists were not too happy with the news that they were cousins to the apes, but the idea that they were the fittest creatures in all nature made up for that. They did not need to lose sleep over the poverty and toil they were forcing on their workers in factories and colonies, for their own riches and comforts were simply proof of their natural fitness. In fact, they took Darwin's theory as evidence that their way of life—industrial competition—was the most natural and the surest way of human progress. The "mechanism" of competitive capitalist society was surely the best possible social mechanism for producing the fittest humans through natural selection.

Not long after the theory of evolution became known, the Russian Revolution produced a new social mechanism known as communism, which was meant to be based on cooperation rather than on competition. Russian scientists rewrote the theory of evolution accordingly, to show that cooperation in nature produced more fit natural creatures than did competition.

Science, it seems, was moving farther away from religion and closer to the politics of men who had wrested social and political power from the church. In both the capitalist and communist worlds, scientists were awarded the status of a secular priesthood supported by governments and rewarded for shaping worldviews that fit the politics of their society. Yet much as they argued the natural advantages of competition on one side and the natural advantages of cooperation on the other, industrialism itself dictated a way of life for bosses and workers on both sides. Industrialism, that is, shaped human habits to its needs, making society itself into a kind of great mechanism.

City transportation systems were built to get workers to and from factories, and education systems were designed to produce the workers. Schools trained children to be on time and to sit still for long hours without talking to their neighbors, doing what they were told even if it was boring, as they would have to in factories when they got older. It was as if children were raw materials put into the school machine and turned out as workers. The clocks and schedules of industrial workers replaced the sun and the weather in telling people when to do what. Government systems got more complicated, more centralized, more organized, to manage society in ways that made industry and the trade of industrial products work smoothly.

Thus, families, schools, hospitals, governments, and other social institutions were run as efficiently as factory machines. The whole way of life became as mechanical as the scientific worldview, and new branches of science—economic science, political science, sociology—were created to design and build the machinery of society, to keep it well oiled and in good repair.

The idea of perfecting humanity, first stated in Plato's worldview, was held throughout the Christian Middle Ages and the Renaissance, and that idea fired the imaginations of the founding fathers of modern science and industry. Now it was put to its greatest test. The modern industrial age was to bring the solution to all problems at last; it was to create perfect order in the lives of individuals and in all society.

The later Greek philosophers and the Christians had sought perfection in the practice of *ethics*—the human pursuit of what is right and good for one and all. But modern science did not concern itself with ethics or with any other human values. Scientists were not interested in what they saw as vague and apparently religious ideas of what is right or wrong, good or bad, which could be argued forever. They saw their job simply as the positivist task of describing natural mechanisms and passing their knowledge on to the engineers who would bring both nature and human society under control with perfectly designed and managed technology.

But neither personal nor social nor economic nor political problems were brought under such control. Science and engineering made great advances in industrial production, in transportation and communication, in medical technology, in weaponry, and finally even in the exploration of space, but industrial nations were at one another's throats in the biggest wars ever fought. Their wealth, moreover, had been gained at the expense of vast numbers of Third World people who remained poor, hungry, ill, uneducated, and without opportunity for anything better. The promised Golden Age of humanity seemed farther away than ever.

Meanwhile, scientists were extending ideas about the mechanisms of evolutionary change to the cosmos as a whole. Astronomers traced the universe back to the Big Bang—the original explosion that created all the cosmos we know as its super-hot energy expanded. Stars and galaxies evolved, but moved ever farther apart, so that the universe as a whole was apparently spreading out and cooling off. The great cosmic machine, the astronomers said, was running down—ever slowing and moving closer to its "heat death," when all order would be lost in the final cold of chaotic nothingness.

However far off this end might be, it was a depressing vision, and scientists offered no salvation from it, no comfort of values or ethics to give life meaning. People began seeing themselves as helplessly trapped inside a cosmic mechanism that would run down no matter what they did. Life simply had no meaning in this coldly scientific worldview, but who could oppose scientific knowledge? Modern philosophies such as existentialism and some schools of modern art reflected the scientists' view of a mechanical universe in which humans are caught without meaning or hope.

Many scientists today still believe firmly in just such a mechanical worldview, but many others now see nature as much more than mechanism. Perhaps the exploding energy of the Big Bang at the beginning of the universe *will* run out like an enormous used-up battery after billions of years have passed—but while it lasts, it lights up our universe and gives it the power to create life from that energy. Those who believe that life is self-creating in a dynamically alive universe rather than a mechanical one also believe that life can create its own meaning and purpose.

In Chapter 5 we spoke of the differences between mechanisms and organisms in connection with the autopoietic definition of life. But to really understand the present scientific debate on whether nature is or is not mechanical, we must go back once again to look at just what we mean by the concept and physical reality of mechanism, and at what role it has played in human history.

15

Less Than Perfect, More Than Machine

While the mechanical worldview and the explosion of technological progress it led to are historically Western developments, their consequences in science, technology, economics, and politics have shaped the course of all humanity. Our invention and use of machines has become the guiding force of our whole species' evolution—we are, for better or worse, technological creatures.

The word "technology" comes from *techne,* which originally meant all art, but it has since come to mean the art of building machines. Machines have given us powers far beyond those of our bodies, and we probably began inventing them to compensate for body parts that we lacked, such as long teeth, claws, and fur.

Our earliest machines—designed to extend the power of our hands and arms—were levers to move rocks, slings to throw stones, bows to fire arrows. To feed and clothe ourselves we formed flat and hollow stones for grinding and pounding food, spindles and simple looms for making cloth. All these are machines, as distinguished from tools, in that they have parts which move in relation to one another. As our civilizations developed, we invented winding mechanisms and wheels, nuts and bolts, pulleys, and other ingenious devices for improving our machinery. We built mills and carriages and great machines of war to hurl missiles at enemies and climb their walls. But the real explosion of human technology came much later with inventions such as the printing press and the spinning jenny, which made useful things in larger quantities and less time than ever before; with inventions such as steamships and locomotives, which moved people about in larger numbers at greater speed than ever before; with inventions such as the radio and the telephone, which let more people communicate farther and faster than ever before.

Machines are made of parts that move to do something humans wish to do. In the first machines the parts were moved by people themselves or by domesticated animals, so it is easy to see them as extensions of people. But as water power, steam power, fossil fuels, electricity, and finally atomic power were harnessed to machines, they seemed to take on a life of their own and we tend to forget that machines are still now, as they always were, a part of humanity invented by humans to extend human powers.

Mechanisms come into being and function only through human design, manufacture, and use. They extend our power to build, to make things, to go places, to fight wars, to measure time and space, to perceive much more of our world and our whole cosmos in its tiniest and vastest reaches than can our senses alone. Machines extend our power to amuse, teach, and talk to one another to show ourselves to one another around our whole planet. They even extend our power to remember, to think, to predict and plan our future.

In all these ways and more, machines extend the powers of their designers and users. No machine would ever have existed without a designer and builder, not even the automatic machines that seem most independent of us. Science fiction writers may imagine worlds run by self-designed and self-reproducing machines, but machines will never exist without human creators and users somewhere in the background.

The idea that computer-run robots could come alive on their own is part of the misunderstanding even scientists have of mechanisms. Those who believe that life evolved by accident in a mechanical universe, on a nonliving planet, can also believe in accidents that will make robots come alive. But the fundamental distinctions between living organisms and machines show us why this will never be so.

Let us review those distinctions. Living organisms or systems remain functional only by continual change, whereas mechanisms remain functional only if they do not change. (Note that changes in natural systems can progress in only one direction, as they cannot undo their aging, but machines can run, in principle at least, both forward and backwards.) Living organisms are autopoietic and autonomous—that is, self-produced and self-ruled—whereas mechanisms are, to coin the appropriate word, *heteropoietic* and heteronomous—other-produced and other-ruled. The "others" are humans, or human-programmed robots, which make other robots. A robot making itself by its own rules is a logical impossibility.

○ ○ ○

If we understand machines as extensions of ourselves and then think back to the mechanical worldview of Descartes, we can see that it was at least logical, for Descartes understood natural mechanisms as God's creations—as engineered extensions of God's power. It was later, when scientists decided to explain nature as a self-evolving mechanism, without any creator, that a contradiction arose. Scientists compared "natural mechanisms"—which they believed to exist without purpose or design—with man-made mechanisms that do exist on purpose and by design, and which cannot exist without creators to design them for their purposes.

Take, for instance, the human brain. When scientists took God out of their worldview, they also had to take out the idea of the human mind as a copy of God's. That left mind as a mechanism itself, or as the work of the brain mechanism. But just what kind of a mechanism could the brain-mind *be*? Some scientists saw it as a plumbing system of pipes and valves through which thoughts, feelings, and instincts flowed like water or got shut off and built up pressures that caused problems. When telephones were invented, the brain seemed more like a telephone exchange of messages along nerve-wires. No sooner were computers invented than the brain seemed to be a computer. More recently, some scientists have chosen to see it as a holographic camera and projector, which are among their newest inventions.

Now, in some ways all these ideas were and are useful, for each of these man-made mechanisms—the plumbing system, the telephone exchange, the computer, and the holographic camera and projector—could be taken as a model of some aspect of the brain-mind in a way that would help us understand something about it. There is nothing wrong with using our mechanisms as models of nature as long as we remember that they are only models and that they can only model certain measurable aspects of things found in nature. The contradiction arises only when scientists confuse the model with the reality they are studying—

when they believe, for example, that brains *are* complicated computers just a bit more sophisticated than present man-made ones, rather than seeing that computers are simply useful models of certain limited things brains *do*. Computers do these things in entirely different ways from brains, yet the model of the function can be valid as such.

The confusion of models with reality comes from a failure to understand that scientists create *abstractions* in quite the same way that artists do. If they did understand their models of nature as abstractions, they would no more confuse those models with reality than artists confuse their paintings or sculptures with the real subjects they portray.

But just what is an abstraction? To abstract means to lift out or away from. An artist lifts out certain perceptions of something and makes them into a painting, while a scientist lifts out certain measurements of something and makes them into a scientific model. In both cases, the painting and the model are abstractions that stand for the whole thing. A mechanical bird can be considered an abstraction of a live bird into an assemblage of metal parts, just as Picasso's *Guernica* is an abstraction of human warfare into an assemblage of brushstrokes. Similarly, a scientific computer model of a biological or economic situation is an abstraction of certain measurable elements from the real biological or economic situation.

Now let us reconsider the mechanical worldview of Descartes as an abstract world model. Descartes abstracted just those measurable features of nature that men were able to copy in mechanisms such as wind-up birds and church-tower puppets—call them the mechanical aspects—and let them stand for all nature. The creative essence of life, on the other hand, he abstracted from nature and put into God. Instead of seeing nature as autopoietic, that is, he saw it as God's heteropoietic creation.

Now we can see that the real danger of confusing scientific models with reality is that aspects of nature which we cannot measure, and therefore cannot abstract, may be the most essential aspects there are. Descartes's worldview, or world model, was at least logical, since he understood that mechanical nature could not exist without an Engineer. But later scientists who dropped God from their explanations of nature failed to see that they were dropping the very essence of life from their world model. Much as they have tried to explain life in mechanical terms, their explanations have never been satisfying.

Scientists who do not mistake their models for reality—readily admitting they are only models—may still consider nature entirely mechanical by arguing that it is far more complex than, though in essence the same as, present mechanisms. But more and more scientists are dissatisfied with the mechanical worldview, recognizing that it is the self-creative aspect of nature that none of our mechanical models can account for. They are coming to realize that nature must be far more than mere mechanism, that it has a creative aspect no machinery can have.

The mechanical worldview originated with those ancient Greeks who decided that geometry was not a human invention, but the human mind's recognition of nature's true design. Yet however much they and later scientists believed in a natural geometry, no one has ever been able to squeeze nature into a neatly geometric or mechanical form. It has proved impossible, for example, to fit any of the cycles of sun, moon, earth, or other planets into the perfect geometric cycles the ancient Greeks claimed were their true motion. Every calendar devised by humans has been plagued by the irregularities of nature.

Geometry would never work if its rules were as flexible as those of nature. On the other hand, nature would never work if its rules were as *inflexible* as those of geometry! Nature, we might say, is more an artist than an engineer. Nature uses the same materials and the same schemes again and again, but endlessly creates something new from them and never machine-copies anything.

When we humans express ourselves through our technology, we usually copy some part of nature that we can abstract and translate from nature's evolutionary artistry into our static engineering. But we must remember that our human ability to copy some aspects of nature in mechanical form does not in any way prove that nature itself is mechanical.

Let us go back to our earlier discussion of thermostats. The thermostat we install in a house so it will keep itself at the same temperature is a mechanical device designed to simulate what every warm-blooded creature does—and what our whole living planet does. But the "thermostats" in our bodies and in the earth are vastly more complex. Such natural thermostats cannot be removed from their living bodies, reduced to their parts, and rebuilt. They exist only in place as a feature of the whole body, and we can only search the intact body for aspects of thermostasis such as vasodilation and sweating.

We have copied the spinning and weaving of spiders, the termites' building of very tall structures, the trees' pumping of water against the pull of gravity, the tunneling of creatures into the earth and their flying into the sky. We have copied the ability to see through darkness and detect things by sonar, to produce chemicals, solar power, and by now almost countless other natural wonders including the ability of our brains to solve problems. But though we can make mechanisms to copy things other creatures do, we cannot even come close to building a working mechanical copy of the simplest single-celled creature as a whole. Our mechanics are limited in ways that nature's organics are not.

Does this mean that we must abandon mechanical models in science in order to understand nature? Not at all. We said earlier that the only way we can ever understand *anything* is by comparing things we don't yet understand with things we do—things that are familiar to us. And what can be more familiar to us than the things we ourselves have designed and created? If we had not invented the mechanical worldview along with our other mechanical inventions, we might not have made so much progress in understanding our world. But we must keep our minds open and recognize that nature is far more than mechanism, that we will hold up further scientific progress if we mistake our present models of nature for nature itself.

○ ○ ○

Descartes and the great physicist Newton built their worldview into a frame of space and time. Space and time were believed to have existed before the universe came into being, as a kind of stage on which atoms and the larger bodies they formed had been created and moved about lawfully. Each atom had a very definite location in space at any given time and moved to new locations as time passed, according to fixed laws of motion. As the French astronomer-mathematician Pierre Simon de Laplace put it, an intellect that knew the positions of all the atoms in the universe at any given time could predict the entire future of the universe.

During the nineteenth century this simplistic model was shaken by the discovery of electromagnetism and the new science of heat—thermodynamics—which came to be called the science of complexity. But perhaps the greatest blow to the mechanism analogy came when physicists were finally able to study the atom itself.

All atoms, remember, were supposed to be exactly alike, though they formed all the different things found in nature by being arranged in different patterns. Far too small to be seen, they were believed to be so hard they could never be broken or destroyed—they were thought to be not only invisible, that is, but also indivisible.

Even if nonmaterial electromagnetism had to be added to mechanical nature, even if heat made things seem to behave erratically, the ability to study the atom itself would surely confirm the mechanical worldview. Atoms, the smallest parts or building blocks of natural mechanism, must be moving lawfully in time and space. Scientists were at last ready to work out just how things were built from the bottom up.

Or were they? We already mentioned the first shocking surprise they got—the realization that atoms were not all alike and were not tiny hard bits at all. Each atom seemed to be more like a tiny solar system, though its shape had to be guessed at from how it acted together with other atoms in forming chemicals. Scientists often have to work this way to figure out the shape of things they cannot see with their own eyes. Think about our solar system—instead of being too small to see, as atoms are, it is much too large for anyone to see all at once. Its shape had to be figured out from the way the parts we *can* see act in relation to one another.

Anyway, just as the sun is at the center of our solar system, something was at the center of the atom, with smaller things apparently whirling about it like planets. Physicists called the center of the atom the "nucleus" and the things whirling around it "electrons." Apparently, different kinds of atoms had different numbers of these electrons in orbit at various distances away from the nucleus.

The next surprise was that the atom's nucleus was itself made of parts, more tiny bits held together by forces so unbelievably strong that splitting the nucleus into its separate parts made an explosion. We all know what that discovery led to.

Every atom, no matter how tightly it is locked into its place—as, for example, in a crystal—turned out to be a tiny mass of jiggling, whirling parts. All the parts, around the nucleus and inside it, are nowadays called particles. But these particles soon proved not to be solid things either.

Deep in the very heart of matter, we now know, there is

nothing solid at all. Particles are like tiny whirling winds in a storm of energy, or like waves dancing on a sea of energy. When physicists try to catch hold of them they rush off, leaving pretty curved trails. They disappear, divide, merge into one another, and reappear out of nothing—in fact they do anything but hold still to be studied. All the physicists can describe, or try to describe, is the pattern of their energetic whirlwind dance with one another—a dance that is, in fact, made of pure energy.

Such discoveries truly confused physicists, whose view of neatly ordered mechanical reality was shattered by them. Particles were neither solid nor reliable; they could pop in and out of existence with alarming speed and mystery. Einstein, furthermore, showed that time and space did not exist by themselves, as a stage for natural mechanisms, but were two aspects of the same concept and were really relationships created continuously as the universe created itself. Space did not even obey the laws of Euclid's geometry as had been believed, nor did time tick away in one great perfect clock rhythm. Instead, it seemed that cosmic space-time curved like the dances of its tiniest particles, and one man could travel through this time-space without aging while his twin stayed home and became an old man.

The world seemed to dissolve at its very foundation. Yet it didn't dissolve into nothing, for the moving pattern of the energy dance is always there giving matter its form. It's just that separating the dancers from their dance—to study a particle as an object in itself—is quite impossible. As impossible as trying to take winds out of the air and waves out of the sea in order to study and understand a storm. If you try, you will find you have nothing at all in your hand—even though you know the storm is made of wind and waves.

The universe, after all, cannot be separated into parts as can a machine. Physicists have to be ever more inventive to learn about their strange new universe. They study particles, for example, in cyclotrons—the largest machines ever made, designed to allow scientists to study the tiniest "things." But to see even traces of the particle dance, they must disturb it and try to work out what the real dance is like from traces of this disturbance. What matters, it turns out, is the pattern of the steps in the dance, for certain patterns of energy *are* what we call "matter." Dancers not dancing are no dance—and the dance is all there is!

Though we can never see the natural particle dance undisturbed, we can be sure it is there—forming and connecting the stars and their reflections in the sea, the earth and all its creatures, ourselves and all the things we make and use. Everything is made of countless invisible dancers' movements in one single dance forming endlessly new patterns—a dance far too small to see and yet so large that it *is* the whole universe.

These discoveries, together with physicists' discoveries of larger dancelike patterns—patterns of wave mechanics in gases, liquids, and solids; of thermodynamics in heated matter; of electromagnetism—called for new ways of modeling nature, new kinds of mathematics that are less mechanical, more flexible, more like living nature.

All mathematics, up to the present, has been built on a foundation of mechanics devised at about the same time by Aristotle for logic and by Euclid for geometry. Yet, until recently, the connection between logic and mathematics was not obvious even to mathematicians. Now that it is, mathematicians recognize logic—rules of orderly classification and combination of elements—as the true foundation of mathematics. And what this means is that mathematics can be changed as much as worldviews, because its logical rules can be changed.

So far, mathematicians have just begun to fiddle with Aristotle's logical mechanics, and already new and more dynamic systems of mathematics have been and are being built on these changed foundations. In time, new kinds of logic, new ways of ordering human thought about a dynamically alive universe—an organic rather than a mechanical universe—may lead to a whole new kind of mathematics that will be useful in modeling the universe. Science and mathematics are now working hand in hand on their exciting co-evolution.

Among the most important new studies in science are those aimed at understanding the self-organization of natural systems. Many of them have been inspired by the work of Nobel Prize-winning chemist-physicist Ilya Prigogine, who revived the ancient concept of nature's creation of order from chaos, showing how self-maintaining systems even at a chemical level can re-create new order when they reach chaotic states. Prigogine's work extends the physics of equilibrium thermodynamics, which was invented to describe nonliving systems, into non-equilibrium thermodynamics, which he used to model living systems. Yet if it is true that living systems are fundamentally different from the mechanical conceptions of physics, then we must question whether non-equilibrium thermodynamics could adequately describe the autopoietic process we have taken as the essence of life in this book.

○ ○ ○

Such new theories and questions are part of our rapidly evolving scientific worldview. New as it is, on one hand, the organic worldview is, on the other, even more in tune with worldviews held by the earliest Greek philosophers, before the concept of natural self-creation was suppressed and God's perfect order became such an obsession that the whole Western worldview was changed to fit it.

Today's scientists are discovering things that people who were very close to nature understood in their essence, if not in scientific detail—as people who are still close to nature feel and tell them in the symbols of dance and myth. Most important, scientists are beginning to understand the universe as alive and ever changing, as a creative dance of life.

The ancient Greek myth told of Gaia's dance; the Indian myth told of Shiva and his wife Shakti, who forever dance the universe and our world into being. Of all creation myths, none tells of a world assembling itself mechanically as tiny parts come together to form the larger parts, which then come together as a whole world—none but that of mechanical science. Rather, most creation myths begin with a whole—a simple wave or whirl, or a single being, that divides into or gives birth to the different things we see as parts of the world. These parts may later rejoin as new wholes—or holons—within the great dance holarchy.

We have seen that living systems are in many ways the antithesis of machinery; we have seen that images of dance fit many aspects of our new understanding of nature better than mechanical images do, for dance is a living, self-creative process as is nature in evolution. Consider once more the way we create a dance ourselves. It may well begin spontaneously, as a natural expression of our energy that is not planned or designed in advance—as an improvisation, in other words. Yet we also speak of a dance's evolution as new variations on the same basic steps create ever more intricate and meaningful patterns—which is just what happens in natural evolution.

Some of the patterns in human dance may even be quite mechanical, as we express our mechanical ability through them, though we see and feel that the more mechanically perfect they are, the less lifelike they are. Classical ballet was an intentional effort to make dance as perfect as possible, and it developed at the same time our machine age developed. Some ballets appear to be danced by robots, rather than by people. These are not considered very good ballets, even if they are impressive. The art of dance seems to depend on human variation, on personal style, on imperfections, on surprise, to give it life and interest.

Nowadays, classical ballet is far less popular than dances with freer patterns, and this may well be because our human search for perfection in the world and in ourselves no longer fascinates us as it did during the past few thousand years. We seem to have satisfied our longing for perfection in building close-to-perfect machines. We want such machines to free us from our own boring, mechanical tasks, but we don't want our society to treat *us* like machines or like machine parts. We are tired of being told to be in perfect shape, in perfect control of our lives, because we begin to see now just how unnatural perfection is.

Nature is orderly without being perfect, as we have seen again and again. Nature's most useful patterns are never outdated but are kept for endless reuse, and the overall scheme of evolution is very stable and resilient. But mechanical perfection would be death to nature as it would be to us as part of nature, for nature is a live, self-creating process forever making order from chaos, forever free to do something new—to reorganize itself when necessary, even if only to stay the same, to create new forms when old ones no longer work. Perfection would be the end of evolution, the end of freedom, the end of creativity. We have learned that nature is far less than perfect for a very good reason—for the same reason that nature is far more than mechanism!

16

The Body of Humanity

O The new scientific worldview we are forming is already showing great influence on our broader cultural worldview. Just as mechanical images inspired the development of industrial and social technology, organic images of self-creating networks are beginning to inspire us to reorganize all human society as a more harmonious and humane venture.

Gaian evolution itself is pushing us in this direction. The evolution of a worldwide body of humanity is very much a part—in fact, the newest part—of Gaian evolution. And like the rest of evolution, it was not planned. Much as we humans have been creating it through our technology, we have not been creating it intentionally any more than we intended to destroy the environment as we created our industrial lifestyle, or any more than we set out to create a means of committing species suicide when we invented nuclear weapons. We have just begun to understand that these are the consequences of recent human activity and that they put our very survival in question.

In just the past few hundred years of our few million years as humans, we have used our big brains and clever hands to produce a technology that changed the whole planet and united us into a new kind of being. Yet we have hardly even been aware that it was evolving us into a single body of humanity. It happened as naturally as the evolution of our physical bodies. Most of our understanding of ourselves, of our evolution, and of our social history has, after all, been gained only very recently. Before this century, we couldn't even know what was happening in the rest of the world *while* it was happening, much less trace its roots into the dim past. Quite suddenly we live in an age of telescopes that show us the most remote parts of our universe and its most ancient history, an age of microscopes that let us look deep into the tiniest parts of our own bodies and the rest of nature. Only in this age have we begun digging up the fossils of our early ancestors and the remains of the first human civilizations, making them into books and films that tell an ever more connected and meaningful story.

Only now can we see our whole planet from space and begin to understand it as a great living being. Only now do we see that, from Gaia's perspective, life evolves as a whole—rock transforming itself into what we perceive as a great variety of separate species, as well as into what we see as the various environments of land, sea, and air. But, as we have seen, environments are not lifeless geological habitats in which living species evolve; they are themselves collections of living species and their products. Sea, soil, atmosphere, and even hard rock are all products of earth's geobiological metabolism as a live planet.

If all creatures and environments co-evolve by changing themselves and one another, then to understand any particular species we must try to understand how its evolution is related to the evolution of its environment. As we said, rabbits cannot evolve without their rhabitats and vice versa. In particular, we can only understand ourselves as humans by trying to understand our co-evolution with our environment.

Let's go through the evolution of humanity just once more. Let's imagine seeing it from afar and sped up like a short film. First we see small groups of humans evolving from their ancestral apes in dense forests in the warmer areas of the earth. The climate changes and the forests shrink, and we see the evolving humans exchange a life of swinging through trees for one of walking upright on the ground. Groups of them wander in search of food. Slowly they begin making tools and weapons, taming fire, wearing clothes—using the resources of their environment to make things that are of use to them, things that compensate for their lack of fur, sharp claws, and long teeth; things that help them hunt other large animals for food, bone tools, and clothing; things that help them carry, store, and prepare food. When their families or tribes grow too large to live together easily, some members bud off to form new tribes.

The human creatures thrive, multiplying and spreading out, following food and water supplies. Great ice ages squeeze them back toward the equator, but each time the ice thaws, they are lured toward the lush new growth springing up in the wake of the ice. Eventually their food supplies draw them to all the continents, even across the land bridge that is now the Bering Strait, into North and South America.

In the best climates, groups of them settle to make villages and fields, to keep animals and grow crops, to store food for dry or cold seasons. Villages grow into towns, and towns into larger agricultural societies that transform considerable plots of ground from natural to artificial ecosystems. Living in peace and equality for thousands of years, budding off new colonies as they grow and thus alternating with nomadic existence, they spread over the habitable areas of the world, developing their arts of plant selection, animal husbandry, pottery, painting, and metalwork. Then they are suddenly overrun by other humans, tribes of wandering nomads and hunters from harsher climates, armed with weapons, taking them over, establishing a dominance system of males over females, rulers over those ruled. They build kingdoms and unite them into empires through warfare. More and more land is taken for human use. The old self-creating, self-balancing ecosystems are destroyed as natural plants are cut or burned, their animals driven off, both replaced by human-bred crops and livestock, as well as by cities of stone and brick.

Within and between empires, wars are fought and goods traded, building networks of land and sea paths that connect human societies with each other. Along these paths, news, ideas, and stories flow together with people and their products, animals, seeds, microbes. Sometimes unwittingly, people change whole ecosystems as their seeds or animals take over and drive out the native species. Cities, in which natural land is replaced by man-made buildings and streets, grow up as centers of ideas, inventions, new ways of life. Their crowded conditions also breed disease; plagues sometimes wipe out whole populations.

The borders around kingdoms and empires change; continents are mapped into countries; human populations grow and divide into ever more languages and cultures. The environment has shaped human civilization by drawing it to favorable climates, into fertile river valleys, and along the easiest overland transportation routes. In turn, humans transform the environment ever more to their use.

Cities crowd more and more people together in artificial environments; raw materials are transported to the city centers from more and more distant places, while the products manufactured from these materials flow outward again toward markets. Crops and animals native to one part of the world are planted and raised in others. Human technology evolves from horses and sailing ships to steamships, jet planes, and spacecraft, from weaving looms to computer industries, from town criers to television. A world once dark by night except for forest fires is lit by a twinkling cobweb of electric lights. A world once silent by night except for the lone cry of a bird or mammal is filled with the sounds of machines and music. Mines and quarries have been dug deep into the earth and scratched out of its surface, their stone, metal ores, and fossil fuels transformed into human products. Rivers have been dammed up and diverted into unnatural paths, flooding ecosystems behind them, making deserts in front of them, for the sake of the insatiable human demand for electrical power. Whole forests have been cut for lumber and fuel or burned to clear land for grazing and agriculture. More and more natural land is plowed under by farmers and paved over by builders of cities. Deserts grow larger while more and more species of animals and plants are killed off as humans exploit nature for their own purposes.

The atmosphere, the waterways, the soil, and the oceans become polluted by man-made fertilizers, pesticides, heavy metals, and other waste materials of human production. Yet health discoveries have exploded humanity itself into such numbers that it seems to humans they are running out of space—that the earth is bursting at its seams. Deserts that were created by man are later flooded by man in the hope of extending agriculture; in a few years they dry up again.

Nuclear energy is discovered; two atomic bombs are blown up to deliberately destroy man's own artificial ecosystems; others are blown up just as tests, destroying natural ecosystems, raining fallout from the atmosphere worldwide. Human technology makes the leap into space, and humanity, for the first time ever, sees its exquisitely lovely planet from afar, as a living whole. Humanity suddenly awakens to the recognition of the vast damage it has done to its environment, begins to fear the exhaustion or irreversible pollution of natural waters, fossil fuels, and other supplies, to recognize its power to destroy the whole human world and force the planet into new paths of evolution, to feel the effects of its greenhouse gases in an atmosphere that is growing uncomfortably warm, threatening the end of our species in yet another way we have paved.

○ ○ ○

It is an impressive scenario, but the saga ends on a frightening note. One species—new, an upstart—has appropriated the entire planet to itself, turning rich and varied ecosystems into fragile monocultures, vast deserts, and choking pollution. Is it a kind of planetary cancer, looking heedlessly to its own expansion at the expense of its own support system? Why is the only species with so much hindsight and foresight so destructive?

The most obvious feature of human social, political, and economic systems continues to be empire building through dominance: the female half of the species is still largely under the control of and exploited by the male half, most of the earth's countries are still dominated and exploited by the few most powerful ones, individual countries maintain their own dominance systems of class, caste, and discrimination, the few hiring the many to work for them and bring them profits.

Wars have continued to erupt over land, resources, and beliefs and are fought on an ever larger scale, ever farther from home, and involving greater numbers of people spending longer times in strange countries. Some stay after the wars, making friends with their former enemies, marrying and raising children together, while others leave their war-torn countries to seek a better living elsewhere.

More and more people in peacetime swell transportation systems as they are sent to work and live in one another's countries, or as they choose to go there on holiday. They learn one another's ways, sharing more and more ideas. Cultures are mixed within political borders; cultures are shared through networks of local and foreign communication; ever larger numbers of people become literate and learn what is happening in their world. Even people who never set foot in another country can eat and use the whole world's products and know the whole world's ways of life in full sound and color.

People prove that they are capable of mingling and sharing, yet governments maintain hostilities. Artists and scientists try to bridge the gaps between hostile peoples, to share their works and knowledge, their fears and hopes for humanity. The threat of nuclear war has driven even politicians to seek new ways of working out differences. The old separations of distance, language, and culture are bridged as the human technologies of transport and communications bind humanity inevitably into a single worldwide body.

We have already compared the evolution of the "body" of humanity with similar events earlier in our planet's evolution, suggesting that the development of communications and transport and the shift from competitive exploitation to a cooperative division of labor are comparable to earlier processes as ancient bacteria evolved into protists, as protists evolved into multicelled creatures and as modern countries evolve into a worldwide body of humanity. By this comparison, the body of humanity is still not fully evolved, as exploitation still exists side by side with cooperation, and the vital importance of diversity to divisions of labor has yet to be recognized. However impressive our progress in cross-cultural communication and cooperation, countries still exploit, attack, and threaten each other to the point of endangering our very survival as a species.

Very likely there were far more failures than successes as ancient bacteria evolved into protists, instances in which unceasing exploitation and hostilities among bacteria multiplying within a single cell wall led to the destruction of the whole enterprise. We humans cannot afford such failure, as we have only one chance. The common cell wall that binds us together is the boundary of our planet itself.

If we understand the evolutionary pressure on us now to complete the organization of this new body, we can work at the task consciously and rapidly. To see more clearly what needs to be done to complete our organization into a single healthy organism, let us look again at the successful evolutionary precedents: eukaryote cells and multicelled creatures.

The organization of the bodies of multicelled creatures such as ourselves is much like the organization of eukaryote cells, except that organs take the place of organelles, a brain evolves instead of a nucleus, flowing blood and lymph instead of flowing cytoplasm acts as a transport medium for supplies and wastes, and so on. Since most of us are more familiar with the workings of our own bodies than with the workings of single cells, it may be easier to see the relationship between our individual bodies and the whole body of humanity than to keep talking of cells.

Let us, then, play out this metaphor, or analogy—this comparison of our familiar bodies with the still unfamiliar great body into which we are uniting—by regarding countries as organs, by seeing shipping routes that carry supplies and products as blood and lymph systems; communications networks that spread information and ideas as a nervous system; and attempts at building some kind of world government as first steps in the evolution of a brain that can coordinate all the body's activities. And let us acknowledge that the Gaian experience accumulated in the evolution of our bodies, as well-functioning and representative living systems, is worthy of respect.

Consider economics and politics—the ways in which we manage our products and ourselves. How do we organize these basic functions of collective humanity and how does this compare with the organization of basic functions in our physical bodies?

Economy—the way we organize the making and shipping, the buying and selling of our human products and services—meant "rules of housekeeping" back when the word was coined and everything people ate and used was grown or made within households. Our human "household" is now all the earth, and our economics is a worldwide system of manufacture and trade that works by both national and international rules. Yet this system did not evolve to serve a worldwide household at all—it was not intended as one great system. It grew out of rather lawless competition among individual nations, though it was forced by its own evolution to make international rules for managing it. Unfortunately, these rules still serve the interests of competitive exploiters better than they serve those who are exploited.

The industrial countries that set up the international economy simply have more money and power to make political and economic decisions than do the poorer countries that supply their raw materials. If we continue the analogy with our own bodies, we can easily see why this is an unhealthy situation. The parts of our bodies—its "nations"—work together as organs and organ systems, such as bone, blood, muscle, and digestive organ systems. If all these organs and systems did not work harmoniously within themselves and with one another, our bodies couldn't function.

Imagine, for example, what might happen to us if our bodies' food economics worked like the food economics of human society. Blood cells are produced inside bones all over the body, just as raw materials are produced in supplier countries all over our world. The heart pumps the blood cells to lungs and digestive organs, where they are transformed into food and oxygen carriers, and then pumps them on throughout the whole body to nourish it.

Suppose, now, that the heart and blood vessels were owned or controlled by the "industrial" lungs and digestive organs.

These organs buy the incoming "raw material" blood, cleanse it of impurities, transform it into nourishing blood, then pump it only to those parts of the body that pay the prices they set.

Imagine the announcement, "Today's body price for blood is such-and-such. Who will buy?" Some of the bones in which the raw material blood cells are produced can't afford as much food and oxygen-rich blood as they need to stay healthy. But rather than lower their prices, the industrial organs destroy the "surplus blood" that no one can afford to buy. Bone cells begin to die of starvation. The starving bones would soon affect the whole body, making it unhealthy, crippling or even killing it.

This is the way we humans manage our economy. Even though our products, including our food, originate all over the world, we do not share fairly the means of their production or their distribution. Our food supplies are presently enough for all humans to eat well, but industrial countries own or control the bulk of food supplies, and they can set prices for the world market. Rather than let prices go down by flooding the market with food, they hoard or destroy surplus food and pay farmers in their own countries to stop producing.

Countries that grow food crops for export to industrial nations often do not grow enough food for themselves, and many have starving populations. Bangladesh and the Sahel countries of North Africa, for example, have suffered starvation during years in which their food exports were maximum. So much of their productive agricultural land is devoted to export crops that the very people who grow these crops have insufficient land for their own crop needs.

It is for reasons such as these that our news media often report starvation side by side with "crises" of overproduction! The solution, except in times of extreme emergency, is not to give away surpluses to the hungry but to redistribute the arable land so that they can feed themselves.

In our bodies, troubles of this kind do not arise, for our bodies evolved cooperative economic systems from the start. Illness or injury can of course stem from outside sources or from internal breakdowns, but our brains quickly detect such problems and see to it that any part in trouble gets immediate help from other parts. In its natural wisdom, our body recognizes that any unhealthy part threatens the health of the whole. It is no doubt fortunate that the freely thinking and acting parts of our brain—the parts that make our conscious decisions—are not in control of such matters, for they have proved, at least so far, much less wise than the automatic, unthinking parts that evolved earlier. We shall return to this observation in the next chapter.

It is obvious that a living body can be healthy only if its systems function cooperatively. As long as human economics remains more competitive than cooperative, we hold up progress toward the evolution of the body of humanity.

The problems in our world economy have become even worse because starving populations—strange as it seems—grow faster than well-fed populations. It is as though the bone cells of our bodies, seeing their kind dying off from starvation, produce ever more of themselves in a frantic attempt to preserve life. Poor people see their children dying and make more. It is natural for them to love their children and to feel that the more they have, the better off they will be. More children mean more workers to bring in family income and care for parents in their old age.

The well-fed people of richer countries do not have such worries, and they have the opportunity to do many interesting things besides raising children. There is no overpopulation problem in rich countries; on the contrary, some rich countries have an underpopulation problem. It has become quite clear that if everyone in the world had plenty of food and opportunity, we would not have developed a population problem. Yet our efforts to solve this problem are all based on curtailing populations by law and contraception rather than by raising living standards and increasing opportunity.

People who are not hungry are also less angry. Much warfare in our modern world is a result of conflict between rich and poor—poor people trying to get back land and resources taken from them in colonial times by industrial countries, industrial countries trying to keep or get back their control over these sources of raw materials.

Which brings us to politics.

○ ○ ○

The most powerful nations have divided themselves into two camps after embracing the competing political-economic systems of capitalism and communism, as we all know. One side says people should look out for their personal interests and the whole society will flourish naturally through their competition, as in Darwin's theory of evolution. The other side says people should cooperate by sacrificing their personal interests to work for the good of the whole society.

The differences between capitalism and communism have actually proved to be a good deal less sharp in practice than in theory. Both systems have essentially the same industrial structure: bosses and workers fill communist as well as capitalist factories and live basically similar lives on both sides. Both sides have recognized to some extent that their own people can no more afford to ignore collective society's interests than collective society can afford to ignore individual interests. Unfortunately, they have not extended this recognition and practice to their international politics.

What their international politics are really all about is the struggle for power—especially the power to control the Third World's cheap raw materials, as Alvin Toffler has pointed out. Each side claims its system would be best for the whole world to adopt and does whatever it can to push or persuade Third World countries onto their side. The competition of the big-power United States and Soviet Union for Third World influence has fanned political conflicts and outbursts of warfare that periodically threaten all humanity with their escalation into global nuclear war. Both major powers have now made enough nuclear weapons to destroy their opponents as well as themselves and seriously damage the rest of nature. Even if they do not use them, more and more other countries, including Third World countries embroiled in such conflicts, are stockpiling their own nuclear weapons. The international trade in arms, especially to the Third World, has become an enormously profitable enterprise.

Yet everyone knows by now that there is no way to fight a nuclear war without bringing on catastrophe for both sides as well as for those not involved in the conflict. If the body of humanity continues evolving, rather than destroying itself, we will see ever more nuclear disarmament agreements between the communist and capitalist worlds. But disarmament will not be enough, for there is another danger perhaps as great as nuclear weapons.

The human mania for making monocultures is apparent in our social behavior as well as in our agriculture, because we simply have not recognized the vital importance of natural variety or diversity in any natural system. No such system or body could function if some of its species or organs had the power to make the other organs over in their own image. Imagine just a single such circumstance—imagine your heart trying to persuade or bully your liver into being just like it. Its success would clearly be a disaster for the body as a whole. Do we *really* want the Russians or the Americans or whoever is not like us to become just like us?

Nature makes it abundantly clear that the secret of success is mutually consistent and cooperative variety.

The fact that humanity divided itself along lines that promote individual versus collective interest, is not so surprising when we look at nature. This conflict between individual and collective interests preceded the organization of protists from individual bacteria and the organization of colonies and multicelled creatures from protists. In fact, the whole Gaian system must constantly work out this conflict.

It is clear that every natural creature from paramecium to plum tree to puma looks out for its own interests by feeding and protecting itself as best it can, just as Darwin said. What Darwin failed to see was that nature is not made just of competing creatures in geological environments, but rather of those living holons we know as wholes in themselves, yet parts of larger holons—all nested into holarchies. Now, if every holon at every level in such an arrangement looks out for itself, we have a situation in which selfishness continually transforms itself into cooperation!

How this is possible can be seen easily by considering our bodies once again, and in the next chapter we will show examples elsewhere in nature. What is important to understand from these examples is that *nature never requires any individual to choose between its own interests and that of its larger body, society, or ecosystem* as humans do in forcing choices between capitalism and communism. Nature works out a balance between self-interest and interest beyond self, as we can easily see in our bodies.

Each of our cells is a living system, or holon, in its own right. Yet, as a holon, it is also a part of larger holons—organ, organ system, and whole body—together forming a physiological holarchy. Clearly every cell manages to look out for its own interests—to care for itself and to reproduce itself—while also working at the interests of its organ, its organ system, its body.

It is no doubt for good reason that every cell in our bodies contains the gene plans or resources for the whole body, since most of our cells must stay in place and thus cannot take what they need from a common gene pool when they need it, as do the streamlined free-flitting bacteria. The genetic directions, or resource libraries, in all our cell nuclei may even be organized as a holarchy of holons representing interests at all levels from the whole body to those of each individual cell. This is speculative, yet we do know that our whole bodies are clones of one cell and that each cell "switches on" the genes that concern its particular organization and work. What is speculative is whether each cell is in some sense directly informed by its nucleus about the rest of the holarchy's needs.

We know that there is communication among cells and that each cell's organization and work are related to that of its organ, its organ system, and the whole body. The entire system unfolded during our embryonic development in such a way that each level of the physiological holarchy from cell to body looks out for its interests, and thus they are pushed or pulled into cooperation.

If every cell in an organ worked for its self-interest, but the organ as a holon did not, the cells would probably kill one another off in competition. Surely they would be disorganized to the point where there was no functional organ. In the same sense, a society in which people looked out only for their individual interests, because they were not asked to do anything in the interest of their collective society, would not *be* a functioning society. This is why capitalist societies do have governments to manage the public interest, to create public works and institutions, to limit free enterprise and tax some of its profits to meet society's needs.

Consider now the opposite situation—where the organ or the society is so powerful a holon that it can demand the complete self-sacrifice of its cells or people in serving its interests. The cells or people thus enslaved would no longer be individuals in their own right. Science fiction writers have tried to imagine humans becoming robot parts of a mechanical society, but people complain bitterly about being "cogs in a wheel," and they stop functioning well. This is why communist countries have discovered they must give people some opportunity to work for their individual interests if their societies are to work as a whole.

Capitalists are right that people must work in their own interests, and communists are right that society must work in its collective interest, but both are wrong in claiming that one or the other will do by itself. In practice they have adopted some of each other's position, and socialist countries are a blend of the two. Yet the big powers persist in their rhetoric of differences, attacking each other on the public stage of international relations, threatening all humanity over differences that are far less important than what they have discovered and practice in common.

○ ○ ○

The body of humanity has not yet evolved the truly impartial and cooperative world government it needs to coordinate its interests as a whole. Looking back at evolution again, we recognize that there must have been a number of steps in the transition from monera to protists as competition among individuals gave way to their cooperation as members of a new whole. We know that one of the most important steps was the formation of the protist nucleus from the DNA of the various monera living within the same cell walls—the nucleus that could coordinate the information needed to carry out the activities of the whole. The same step was accomplished when nervous systems formed in multicelled animals that had evolved from protist colonies in which different member cells did different jobs.

Something of this ilk is clearly happening as the body of humanity struggles to form its new identity. Since the close of World War I, people have recognized the need for some kind of organization to coordinate and balance national and international interests. First they tried the League of Nations, then the United Nations. Although the U.N. accomplished much in the way of programs and services, the competitive interests of member nations still dominate on important issues, limiting the powers and often preventing the smooth functioning of the U.N. If our human civilization is to survive, we have no choice but to solve this problem before long, completing our evolution into a worldwide body of humanity with a functional coordination system.

Even if we solve this problem, however, there is another problem that is even more basic: How can the body of humanity function if half of its cells suppress the full expression of the other half? It is a blight on humanity that neither the U.N. nor any single country in the entire world has yet paid more than lip service to training and selecting women for half of its governing and professional positions. Nowhere is it recognized that such equality may be fundamentally necessary to the health of any society, that a system of sexual inequality inevitably breeds conflict while losing valuable resources and justifying every other form of inequality, oppression, bigotry, and antagonism.

Our biggest job is to change our whole way of thinking to a larger perspective, to recognize ourselves as a body of humanity embedded in, and with much to learn from, our living parent planet, which is all we have to sustain us. How can we as a species live in harmony within it? How can we as people live in harmony within our own species?

The sooner we recognize ourselves as being in transition from exploitative and divisive practices on all fronts to a united and harmonious living system, and the sooner we recognize that there are natural models to guide us, the sooner we will complete our healthy evolution by our own choice and efforts.

A Matter of Maturation

We have seen human worldviews change dramatically from the early view of nature as the Great Mother to a view of nature as the mechanical creation of a Father God to the portrayal of nature as a mechanism evolving by accident, without purpose or design, there to be used for human purposes.

A psychologist might see this sequence of worldviews that is the heritage of industrialized humanity as having something in common with those of an individual member of modern society passing through stages of personal development. Technological culture, on which this book is focused, has clearly become—for better or for worse—the dominant human culture that will make or break us as a species. What matters, then, is to recognize that this culture is still immature from a developmental perspective. We may then hope it will learn not just from its own experience but also from that of the few remaining nontechnological cultures it is wiping out in its drive to so-called progress, though they may be far more mature in their relation to our parent planet.

Of our infancy we know very little; our earliest artifacts, as we said in earlier chapters, indicate a recognition of nature as mother and of our closeness to and respect for her. In our long, peaceful childhood we learned nurture from her, developing agriculture and art. Father figures existed primarily *in absentia* from these early civilizations, as the gods of nomads and hunters who eventually came to disrupt them with violent conquest and lasting domination.

Under the influence of paternal gods, we formed our *ego*—the Greek word for "I"—coming to see ourselves as separate from nature, growing out of the close union with nature as mother into the objective world where we found security in the laws and geometric creation of this authoritarian Father God.

In this analogy, the European Middle Ages may be seen as the phase of prepubescence—the stable God-fearing Christian society that lasted over a thousand years under God's authority, until man began expressing his ego more boldly—challenging religious "facts" about the world with scientific observations, making discoveries and developing technology in ways that permitted him to transform nature ever more to his own purposes. Was it not as if humanity through its Renaissance and Enlightenment had reached the stage of competence and confidence we associate with adolescence? Science and technological industry swelled the adolescent ego to the point where the father's authority was rejected and nature was seen as no more than a knowable and predictable environment for men to control and exploit as they wished.

Is it not to be expected that a smart and clever adolescent will reject or at least question the unchallenged parental authority in which the child believed? Is it not common to gain in adolescence, between bouts of insecurity, the conviction of knowing everything and being in full control? Why shouldn't whole human societies go through the life stages of childhood and adolescence as each individual human does? Is not our whole species, quite like every child born to it, still young and free to learn from experience?

In mythology—mythology being very close to psychology—the heroic cycle often represents the life cycle, the youthful stage of which is filled with the adventures of a hero who has left home to face challenges but who finally returns as a mature, wise man. Such myths often portray the hero as a brash youth with the hubris to believe himself as invulnerable and powerful as the gods themselves—and he is invariably punished by a fall, which may be permanent if he does not learn the lesson of false pride. In real life, the adolescent who strikes out with a false sense of maturity, believing he or she knows it all, can be expected to get into some kind of trouble before maturing into an adult. And the adolescence of "civilized" humanity is running true to form. Our view of nature as a mechanism to be exploited fostered great progress in technology, but we made this progress recklessly in our belief that all nature was ours to do with as we pleased. Now we find ourselves punished by the enormous problems we have created along with our modern technology.

Like any adolescent who is suddenly aware of having created a very real life crisis, our species faces a choice—the choice between pursuing our dangerous course to disaster or stopping and trying to find mature solutions to our crises. This choice point is the brink of maturity—the point at which we must decide whether to continue our suicidally competitive economics and politics, our ravaging of the environment, or to change our worldview, our self-image, our goals, and our behavior in accord with our new knowledge of living nature in evolution. It is the point at which we can see our own historical evolution and decide whether to continue opposing it with old hostilities or whether to speed its evolution into a mature cooperative body of humanity by conscious choice.

Growing up is not easy, as we all know. Youth must fall on its face in its ambition, must learn by experience; the hero must be wounded in battle and be knocked down for his hubris, his pretension to godhood. Maturity comes only when youth gains perspective on itself and is at last willing to admit there is still something to learn.

As we have not yet gained this perspective, many of us believe that today's human problems will never be solved, that they have simply gotten too big for solutions of any kind or that, even if we solved them temporarily, human nature cannot itself change and therefore we would just get into the same mess again. This pessimistic view of ourselves as a species reflects the way we feel as individuals whenever we are depressed and our problems seem insoluble.

Hopeless pessimism often comes from lack of perspective. If we look at things narrowly—from within a difficult situation—they may well seem hopeless, but if we manage to step out of our dark hole, so to speak, to gain some perspective on ourselves within it, we may begin to see a way out.

The purpose of this book has been to put our human selves into just this kind of perspective—to see ourselves within the whole evolving world, even within the whole evolving cosmos. And when we look at things broadly this way, we see that the problems we humans have created may not be as great as problems other species have created—and which did get solved. What could be more interesting, more exciting, than to be alive in the very age when we as a species have the opportunity to mature, to solve the adolescent problems we have caused ourselves and others?

○ ○ ○

One solution to human problems proposed in the name of ecology is that we should recognize technology as an inhuman disaster, an evil to be rooted out along with the science that produced it, so that we may go back to a simple and "natural" life. After all, those humans who never invented technological languages and machinery have also not suffered adolescent crises of their own making. In fact, natives of the Amazon, New Guinea, and other places where people remained close to nature, cannot be said to be immature in the sense that technological humans are. Though they belong to our very young species, they have never rejected the parental status or lessons of nature, and they have thus learned to live in mature ecological balance.

The rest of us—the vast majority of the species—must recognize that our development has taken a different path and that no living system can reverse its evolutionary history. Our technological development is as natural as was our pretechnological infancy, and we cannot turn back. We could, however, move forward with more mature restraint and wisdom.

Fortunately, our industrial technology is already in transition from crude adolescent efforts to more mature sophistication.

In developing our heavy industry and feeding it with raw materials, fuel, and human workers, we have devastated or polluted whole ecosystems, alienated ourselves from our natural origins, formed ourselves into mechanical societies living in concrete jungles. Yet this phase is already in transition to the next. We have begun replacing heavy industry with light, energy sources that can run out with those which won't, industrial cities with spread-out production networks connected through computers.

Many Third World countries could avoid the expensive and exhausting heavy-industry phase almost entirely by jumping directly into the age of electronics with the assistance of the most technologically advanced nations. In fact this is beginning to happen—but unfortunately it is happening in a context of profit motives and competition, leaving beneficiaries worse off than before, while being forced to pay for the "assistance" in political concessions and loyalties. Thus Third World development schemes continue to devastate ecosystems, as the World Bank now openly admits, and often help promote a climate of threatened, if not actual, hostility. Notorious are the post-colonial "banana republics"—single-crop economies that have created unstable ecosystems and dictatorships that keep their own people poor and hungry.

The building of dams to generate electricity, the burning and bulldozing of forests for monoculture crops or grazing cattle, the use of chemical fertilizers and pesticides, the manufacture of fuels and metals from fossil and ore deposits, are all financially profitable to those who own or rule the land, but so ecologically destructive that the Gaian life system may have to make life intolerable for humans in order to survive itself. Nature works not for profits but for balance. It recycles everything. And humans cannot much longer run their profit economy at the expense of planetary economics, as Thomas Berry, the economist and philosopher priest, has warned.

If we are willing to see the problems we now face as those of a species on the brink of maturity, yet still in the fiercely competitive phase of belligerent adolescence, then we can learn as a species from our individual experience in maturation. Let us recall the classic joke about the young man who comes home again after having struck out to make his way in the world. He is shocked yet pleasantly surprised to see how much his father has learned in the few years of his absence, how much they can at last agree on. The joke, of course, is that the son has changed and not the father—his own experience of the world having given him new perspective on his father as a wise and sensible man with something worthwhile to say.

This is exactly what can happen to us as a species. As a result of our own experience and our recognition of the environmental trouble we have caused, we can take a new look at the planet that gave us birth and we can begin seeing it in a new light. Through our brilliant science, our measuring and computing instruments, our space technology, we can see our planet as a whole living being that we have misunderstood and mistreated at our own expense. We begin to understand that while the planet has great experience and wisdom to teach us, our own lack of understanding and respect have led us to exploit it as though it existed for no other purpose. Only now do we understand that we have been recklessly destroying the parent planet on which we depend and from which we can learn a great deal about using our gift of conscious freedom more wisely.

○ ○ ○

In discussing the problems of free and conscious behavioral choice in earlier chapters, we spoke of our lack of innate rules for dividing land and resources fairly, for avoiding killing our kind, and for governing ourselves peacefully and cooperatively. We suggested that guidance in achieving these things by our own choice could be found in the organization and functioning of natural holarchies. We also recognized that turning our understanding into real behavior instead of just reasoned ideas requires a lot of responsible effort.

The Athenian democracy in ancient Greece was just such an effort, yet it seems to have collapsed at least as much from internal weakness—from the human reluctance to accept the hard responsibilities of freedom—as from external causes. Modern so-called democracies may work as well as they do only because they are much less demanding of their citizens than was ancient Athens—and this is important food for thought. In the United States today, compulsory taxation is the only citizen requirement, and compulsory taxation has never made a democracy.

Late in the nineteenth century, the great Russian writer Feodor Dostoevsky identified the problem of accepting responsibility for freedom as humanity's essential crisis. Dostoevsky presents the crisis of human freedom in a myth within his novel *The Brothers Karamazov,* by having one brother tell it to another. In this myth, Christ reappears in sixteenth-century Seville at the time when the Holy Christian Fathers, in that city alone, were actually burning as many as a hundred heretics a day at the stake. A Grand Inquisitor condemns Christ to his second death, this time at the stake, for preaching freedom to mankind—on the grounds that "nothing has ever been more unendurable to man and to human society than freedom."

Men cannot bear, and so do not want, the responsibility of freedom, the Inquisitor claims, and the church has relieved them of that burden. Men will endure slavery for the sake of being fed, and they will be happy only when their rebelliousness is turned to obedience, he insists, for they are sheep who prefer to believe they are free while actually doing as they are told by authorities who give them work, bread, rules to live by, and forgiveness for their sins.

Interestingly, Dostoevsky calls men "unfinished experimental creatures" in this passsage, implying that they are immature as a species. Further, in his allegory, he deliberately pits two heroic images against each other in an analogy of personal adolescent crisis. Christ, on the one hand, represents the call to maturity—

the acceptance of human responsibilities of free choice; the Inquisitor, on the other hand, represents retreat from maturity—the delegation of responsibility to an authority. Thus Dostoevsky portrayed his painful awareness of the human failure to practice responsible freedom.

Young people all over the world today reflect this dilemma in their personal lives, whether or not they are aware that it is as much a species dilemma as a personal one. Some struggle hard to show all humanity how to accept the responsibility of freedom to end war, hunger, and ecological disaster. Others hurt so much from seeing the human failure to solve these problems that they fall into anarchy and depression, believing in no future at all and driving the youth crime and suicide rate to an unprecedented high. The rest, as Dostoevsky despairingly noted, do not concern themselves with such great problems but do what is asked of them so as to make their personal lives as comfortable as possible.

Perhaps most of us search for comfortable security under some authority other than our own because we have been taught to fear failure if we fly in the face of tradition to exercise our own free choice. We have had the goal of perfection held up to us for thousands of years, and we fear failure and disapproval if we stray from whatever path we are told is the right one. Better to look to some authority for guidance, for the ideology, the formula, that will make us and our societies more perfect, than to risk acting on our own imperfect ideas.

For more than two millennia, ever since the ancient Greeks thought up the idea, we have been chasing after perfection. Now we are forced to wonder if we have not crippled ourselves in this chase after a chimera—a foolish and frightful fantasy. Let's reconsider now, in terms of human freedom and responsibility, the happy discovery that the cosmos is not rigidly perfect as Plato thought, but an imperfect creative process much more as Anaximander saw it, with everything forming and re-forming in the never-ending process of making order from chaos.

Ever since Plato, Western worldviews have held up the goal of making ourselves and our societies as perfect as God's creation—as perfect as well-oiled machinery. Only now do we see that despite billions of years of experience, despite the marvelous integrity of life's patterns, things go wrong in Gaian nature, unbalancing the dance here or there. Yet in going wrong, they create pressure to reorganize the dance of life, to try new steps, or new combinations of old steps, and so the imperfection leads to progress. The story of Gaian creation is the story of improvisation wherever and whenever a species or ecosystem became unstable. Yet in nature there is never any break with the past—there is always continuity in the dance. And the dance always works to produce a remarkable integrity and stability. Nature teaches us that order can be maintained through change—even, when necessary, through disastrous change.

When, for example, the dinosaurs were killed off by severely changing conditions due to an accident, the Gaian system continued to create new reptile, bird, and mammal species from the genes left in the small survivors of the disaster. We sometimes think of the great dinosaurs as unsuccessful species because they became extinct. But dinosaurs flowered into a wonderful variety of species including the largest creatures the earth has ever seen, and they flourished for nearly two hundred million years—forty times longer than our few million years as humans. Nor was their extinction their fault.

It is still up to us to prove that the human big-brain experiment is worth the risk, that freedom from innate rules, the conscious freedom to choose, will pay off in creativity that benefits—or at least does not harm—the whole Gaian system. If it doesn't pay off, we, too, will become extinct, more likely by our own doing than by outside forces as in the case of the dinosaurs. We would be wise to remind ourselves often that Gaia's dance will continue with or without us.

○ ○ ○

Nature, as we said before, is far more like a wonderfully resourceful artist than like a grand engineer, more like a mother juggling family needs, economics, and conflicts than like a coldly calculating geometer. In the improvised dance of nature toward order and balance, complexity unfolds, becomes chaotic or fragmented, is reorganized to new unity, then permits new complexity to unfold, new disorder to arise. The Gaian system protects what is stable, yet is ever open to change when instabilities arise, using change to create both new unity and new variety—variety that gives nature, among other things, the resilience to survive disasters.

Every species is different from all the others and every individual is a variation of that species' kind, just as in our bodies every organ is different from all the others and every individual cell a variation of that organ's kind. Machines can be mass-produced to be all alike, but nothing in nature is exactly like anything else. Nowhere is there an ideal form, in Plato's sense.

It is Gaian wisdom to balance variety and use it creatively in forming highly stable ecosystems. The greater this variety is, the more stable the ecosystem is as a whole—certain ecologists have argued the contrary notwithstanding, as the English ecologist Edward Goldsmith has pointed out. We are also discovering that tampering with such systems by introducing a new species that has not been worked into the dance may disrupt it entirely—as in the case of the gypsy moth or Dutch elm disease.

This variety principle holds also for the gene pool of any species. We have learned by hard experience, for example, that our practice of "perfecting" our food crops and domestic animals by breeding out their genetic variety while breeding in the features we like leaves them weak and subject to devastating diseases. When we reduce variety by breeding a particular strain or even cloning a single individual, by replacing natural ecosystems with monocultures on bulldozed land, we risk creating a highly unstable and vulnerable artificial ecosystem.

Human variety, in our physical makeup as well as in our languages and cultures, ideas and lifestyles, is surely equally important to our healthy survival. Yet oddly, while we humans fight for our individual right to be different from others, not to be forced into the same social mold, we cause ourselves a good deal of trouble as a species by thinking there is something wrong with people who are different from us. We discriminate on the basis of color, culture, or belief—convincing ourselves there is something in difference to be hated, feared, ridiculed, or stamped out. Let us hope such prejudice will disappear as we learn more about nature and begin to respect, welcome, appreciate, and love our individual and cultural differences, using them like genetic variety to create new and fruitful combinations.

It has become obvious, for example, that a common human language is essential to communication and cooperation in the newly formed body of humanity—and English appears to be naturally evolving into that common language—but that is no reason to suppress any of the languages of different cultures. It is no problem at all for human children to learn several languages, and the variety of human languages represents a very important variety in human thought and worldviews. Just as we are foolish to breed natural variety out of domestic food plants and animals while killing off one wild species after another, we are foolish to eliminate variety in human language, culture, and thought.

Many natural languages have already become extinct as a result of foreign conquests. Conquerors have killed off conquered peoples and sometimes even punished survivors for speaking their native tongue, as was done in some of the mission schools that children of Native American tribes were forced to attend. Just as we have begun to work at preserving endangered animal and plant species, we should be working to preserve endangered cultures and languages. When we cut tropical forests, we destroy not only vital ecosystems but also the human cultures that are part of them, that cannot survive being transplanted to concrete jungles—cultures from which we can learn a great deal about living in harmonious balance with the rest of nature.

Nature's adaptive ability to change creatively without ever falling back into chaos surely suggests that we humans should give up the idea of finding the ideal economic and political organization or social structure. We must get used to the idea that we cannot make social progress by forcing people to accept either individualistic capitalism or self-sacrificing communism as a social system, or either conservatism or radicalism within such a system. Our basic and natural task is the same as that of any other species—to balance individual good with collective species good, to be conservative with what is healthy in human society while radically changing what is not.

Modern progress reveals itself as a strange concept. As it is, the "progress" of both capitalist and communist "big powers" is based on very ecologically destructive technological systems and the escalation of nuclear weapon production for the sole purpose of threatening each other. Gorbachev's *perestroika* sets sensible human priorities, however, and the U.S.-USSR arms agreements are encouraging, though it is hard to see the withdrawal of a few nuclear weapons as significant while production of new ones continues.

Nature teaches us that evolution depends on competition *and* cooperation, on independence *and* interdependence. Competition and independence are both important to individual survival, while cooperation and interdependence are both important to group, or social, or species survival. Since individuals and their society or species are holons at two levels—or in two layers—of the same holarchy, we see that these levels or layers must achieve mutual consistency in looking out for themselves, working out between themselves a balance of competition and cooperation, of dependence and interdependence.

If we work creatively to maintain this balance between ourselves as individuals and ourselves as societies—local, national,

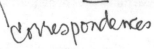
correspondences

and worldwide—we will complete the evolution of a healthy body of humanity. If we look to our own individual bodies as a rough model for making it work, we might see that cooperative peace is a real option for nations with different languages and cultures that can make different contributions to the worldwide economy. There is no reason why individuals should not have the freedom to pursue their own interests and also contribute to their society. There is no reason why all should not be well fed and cared for in an equitable system of work and income.

○ ○ ○

From the Gaian perspective, solutions to the great human problems of war, overpopulation, and hunger are far simpler than from any other perspective. But simplicity does not mean ease. We *could* solve these problems by shifting our worldview from one of international competition to one of international cooperation with the goal of producing a healthy body of humanity. But worldviews, as we said earlier, do not change easily.

The present perspective of the powerful nations—those capable of quickly transforming humanity into a healthy body—is not a Gaian perspective. To attain the Gaian perspective, their leaders must change their worldview and the private or public profit-motivated behavior based on it. They would have to agree to give up the privileged position their nations now occupy and the hostilities they believe are necessary to protect their interests. As it is, these leaders are uneasy at predictions of doomsday, at suggestions that their course is suicidal for all humanity, but they do not yet believe in a healthy alternative that would be as good for themselves as for all humanity.

It is no easier to get the citizens of our modern democratic or communist societies to take on the informed and responsible task of running things, instead of submitting to the oligarchic rule of the few, than it was to get ancient Athenian citizens to do so. But trends toward more Gaian networks in place of top-down authoritarian structures are emerging in new organizations and in various existing industries, services, and other enterprises as they decentralize their management and make it more concentric than hierarchical. As more people come to understand and adopt a Gaian perspective, it is likely this trend will grow.

Just as individuals may grow out of adolescence without consciously assisting their own process of maturation, a cooperative body of humanity may evolve without our intent just by the force of the evolutionary change that is already under way—if we avoid committing species suicide before it can do so. But the process will surely continue with less turmoil and suffering if we stop opposing it and consciously, actively assist it. Would it not be healthier for us to give up our dangerous competition for the greatest financial profits, the most perfect system, and work together at creating an imperfect but ecologically sensible economics and politics, a system that works with its variety to give everyone the opportunity for a healthy life, just as our bodies give that opportunity to each of their cells?

In recognizing our planet as an experienced living system with a good deal of "body wisdom" to teach us, we gain the perspective to see how we might apply some of that wisdom to our own human problems. All over nature, throughout the Gaian life system—right under our noses, so to speak, and all around us—we find the clues to making our own human affairs more organic and ethical, more creative and wise, as the earliest philosophers believed we would.

Let us continue gaining perspective by seeing that great problems can be the very challenges needed to push evolution along into new creativity. We saw that the oxygen crisis of billions of years ago became an opportunity for a new way of life and new forms of living creatures. The environmental crisis that caused the mass extinction of dinosaurs provided the opportunity for our own mammalian evolution. Both the bacterial oxygen crisis and the meteoric dinosaur crisis were more severe from a Gaian perspective than any trouble we humans have caused so far. Yet we are causing enough trouble to call it another Gaian crisis, and it increasingly appears to be a life or death crisis for us as a species. Our increase of greenhouse gases alone may force Gaia to regain stability at an average temperature beyond human tolerance.

Let us give up our chase after perfection and use our differences and imperfections creatively as does the rest of our evolving planet. Let us remember that if nature reached perfection, Gaian creativity would come to an end, for there would be no problems, no challenge, no opportunity for creating anything new. If we follow nature's lead, we will make mistakes, juggle things about, find solutions, generate new problems without guilt, but on the whole we will find our mutual consistency with other natural organisms.

Let us also remember that if we continue on our current path, our planet may be better off without us. Our species death, by suicide or extinction, might actually promote Gaian health. Yet we are potentially as creative as the whole Gaian system we belong to. If we find ourselves in an adolescent crisis of our own making, that is no reason for us to give up in despair. It should, instead, urge us to face ourselves, swallow our foolish pride, adopt a little humility, a wider perspective, and gain mature humanity in the best sense of this word we have coined for ourselves.

18

Ecological Ethics

There is reason enough apart from our youth to adopt a little species humility. We have seen that no species can evolve apart from its co-evolution with all other species. And if we look at the Gaian system through eyes other than our own, we will quickly see that we have no more reason to consider ourselves a supreme form of life than have others.

Back when we spoke of mitochondria and chloroplasts as making up half the weight of all plant and animal cells and being descended from ancient bacteria, we joked about our bodies being giant taxis for bacteria to get around in. Joking aside, the world from a bacterial point of view is indeed arranged nicely for bacterial survival. They live in or buzz in and out of all other forms of life, feeding off the living and the dead, passing around bits of DNA information worldwide. They exist in vastly greater numbers than any other kind of living creature, and there is virtually no place on earth—from the depths of the sea to the highest mountain and the atmosphere itself, from the hottest springs to the coldest glaciers, from the surfaces of other creatures to the depths of their guts—that is not teeming with them. They evolved the whole Gaian system, having it to themselves for most of its life and still maintaining a good deal of its functioning. Bacteria, in fact, are the only creatures that could survive without all the others. Why should bacteria not think—if they could think—that the world is all theirs?

Then, take the funguses—a kingdom of life in themselves. They, too, are spread out almost everywhere, though most are too small or fine-webbed for us to see. Every rooted plant has funguses twined in its roots, bringing it supplies in return for ready-made food. Funguses live on animals as well as on plants. From their point of view, all nature would seem to have been created as their dinner table.

Animals might well look this way upon plants—as though plants had been created especially to feed and serve them. After all, animals eat plants to burn out of them with oxygen the energy they need—and that oxygen was made and put into the atmosphere by the plants themselves, as if to supply animals with breath as well as with food. Animals make use of plants even for their drinking water, lapping or sucking dew, drinking rainwater that was first pumped into the sky through the roots and stems and leaves of plants. Animals also use plants for shelter, making their homes in seaweed and among the branches and roots of land plants from the tiniest club mosses to the tallest trees.

Surely animals could not be blamed for believing that plants had evolved just to provide them with food, oxygen, water, shelter. But what if we shift our perspective to that of plants?

From their own point of view, plants might very well think that animals were created to provide for *them*. Plants—that vast range of photosynthesizers from little more than fancy bacteria to great banyan trees, each a forest in itself—have considerable reason to see themselves as superior creatures. They need not run about chasing after food as animals do, but can sit right where they are, easily making their own food and energy from sunlight and soil chemicals provided by bacteria, funguses, worms, insects, and other animals. The carbon dioxide they use in making energy is also provided by animals. Insects carry pollen from plant to plant, making it easy for plants to reproduce without running about for that purpose, either.

When plants have made their seeds, animals continue to work for them, carrying the seeds about in their feathers and fur. Birds and grazing mammals also eat the plants' fruits and digest them. The animals thus moisten the plants' seeds and wrap them in packets of rich fertilizer, then scatter them in new places to grow. Animals, in fact, do all the running around for plants while plants sit smugly being served all their lives wherever they first took root.

And so any form of Gaian creature—bacteria, funguses, plants, or animals—might find reason to see itself as superior to the others. Even rock, for that matter, could see the whole world as nothing more than its own dance, its endless transformation into living creatures and back into rock. Try for yourself the exercise of looking out over your world and seeing all of it—the landscape, the sea and sky, the creatures, yourself and your fellow humans, their airplanes, their cities, the furniture in your house, this book in your hands—all as no more and no less than rock rearranged.

The continents of the earth are still on the move. Ever since the great single continent, Pangaea, or "All Gaia," broke apart, they have ridden their tectonic plates slowly over the softer mantle beneath them. Africa and South America only separated around the time the dinosaurs disappeared, some sixty million years ago. By moving apart, rock separated its life forms, cutting species members off from one another so they were forced into different lines of evolution. From rock's point of view, it directs even the course of evolution through its motion.

Bacteria have always transported minerals about in quantity, forming the ore veins and even the uranium piles we mine today. Over half a billion years ago tiny marine animals began their endless work of moving calcium about in huge quantities by building it into their shells and depositing it on the sea floor, which later thrusts itself up to become land. In the carboniferous era of the first great forests, plants began their job of burying carbon underground to become coal and oil. And so the earth's rocky crust, itself formerly stardust, has reason to see the bacteria, plants, and animals created in its metabolic dance as its own inventions, meant to serve its needs.

We humans, from all these perspectives, would be considered latecomers, an upstart species, killing or endangering others as we make war on all five kingdoms including our own: bacteria, protists, funguses, plants, and animals. We burn and cut forests, dam up and choke off rivers, create deserts, poison the whole planet's water, air, and soil. And in our unprecedented egotism, we behave this way while declaring ourselves the pinnacle of evolution.

Yet we alone are capable of holding a truly broad worldview that represents the whole of nature and includes all possible points of view in addition to our own. We can—and must—gain enough perspective to see ourselves as one part of a much greater living system, or being, and learn to act accordingly. The body of humanity we have described in its present evolution is a new kind of body and, at the same time, an organ within the Gaian body—the latest organ to evolve within it, one that is only now being tested to see whether it will work. Humanity has woefully little experiential intelligence or wisdom. Its behavior is not programmed by evolution but is up to us—a free choice of conscious ideas, decisions, and practices.

○ ○ ○

Some scientists have come to call human ideas, or pieces of information, *memes*. This name is intended to show that ideas and information are related to genes in the sense that they are a new way for nature, in its human experiment, to expand and carry on the business of forming programs that can be shared by all species members and passed on over generations in new combinations that determine the nature of the species. In this view, we humans are programmed physiologically by our genes and socially by our memes. It is memes—our ideas about ourselves and our world—from which we construct the worldviews that shape our societies.

Our memes are now determining the nature of the new body of humanity. The memes from which we have built our present common worldview include ideas of ourselves as divided into competitive nations—nations or blocs of nations competing for resources and power much as the ancient bacteria competed with one another before they formed a common protist nucleus and thus a protist identity. Such ideas of separate nations in competition must be transformed into ideas of cooperation among varied nations as organs in a single body. Blocs of nations such as NATO and the EEC have already been formed and function like organ systems that carry out different tasks such as defense and economics. These systems, however, are independent of the United Nations and in some senses at cross purposes with it. We still lack the overall perspective in which to form a common plan, though our evolution to date has prepared us to make it. If we transform our memes, our worldview, to a common scheme of voluntary cooperation—an internally creative, autopoietic body of humanity—we will abandon our old ideas of the machinery of society and find new organic ways of reorganizing ourselves. It has already been proposed, for example, that naturally bounded ecological areas such as watersheds make more sense as economic-political units than do our present states or provinces, with their arbitrarily drawn boundaries. The inhabitants of such areas would have natural interests in common. Surely the natural boundaries of geographical features contained our original human societies, which recognized their dependence on nature and did not yet see it as territory to be carved up arbitrarily.

Unfortunately, we seem to have lost some of the most important memes generated in human history. In our capacity to look forward we have often dreamed of a Golden Age to come, yet it seems we have quite forgotten the Golden Age of our past, which Hesiod was the last to recall. The massive upheaval of human society that began six thousand years ago was so thorough in promoting dominance and aggression over partnership and peace that we came to see aggression as our natural heritage. Can we recover the memes of civilized equality and peaceful sharing of wealth that seem to have guided settled human societies during the preceding thirty or so thousand years?

However we draw boundaries and organize ourselves, our new body of humanity must be flexible enough to evolve through still further stages. We can be sure that it will always be imperfect by the old mechanical standards. Whatever social forms it will eventually take, making the body of humanity into a healthy, functioning holon within the Gaian holarchy is the greatest task human consciousness has yet faced.

○ ○ ○

Let's look once more at this matter of conscious thought, as it is directly related to the question of ethics this chapter poses.

The Gaian system as a whole has no conscious mind with which to think about its great complexity or plan its future. It cannot abstract models of itself, nor could it use such models in deciding what to do next. Yet the Gaian system functions intelligently and wisely on its billions of years of experience without conscious thought. It functions, as we said, with the kind of intelligence and wisdom that evolved in our own bodies.

Conscious thought is the newest Gaian experiment, and in Gaian wisdom little has been entrusted to it as yet. Our bodies continue to manage themselves without our conscious help or interference—fortunately so, given the complexity of their functions. Our conscious scientific minds—no matter how impressed we are with our ever growing knowledge—are very far from understanding even a single cell or organ well enough to manage its ordinary daily affairs.

It should come as no great surprise to us that the freedom of conscious decision we *do* have gives us so much anxiety. We look around us and see other species functioning on the whole the way our bodies do, untroubled by questions of whether what they are doing is right or wrong, good or bad. Yet we are stuck with choice-making conscious minds that are an experimental substitute for the innate "evolutionary knowing" of other species, and we must use those minds as best we can to decide how to behave.

It is because of this unprecedented degree of choice that we humans alone must ask ourselves what is right or wrong, good or bad for us to do—personally, socially, and as a species. Modern science, however, has refused to concern itself with such questions on the grounds that they are ethical questions and ethics is the domain of religion, not science. When sixth-century B.C. philosophers suggested that ethical questions can be answered by looking to nature, religion and science had not yet been separated, both being aspects of the same search for orientation and guidance. Only much later, when they were separated by modern scientists, did ethics become the domain of religion while scientists insisted on their ethical neutrality, their freedom from values.

If the ancients were right—as this book holds—that nature is indeed the best source of guidance available to humans, then surely science, as the study of nature, is the most appropriate field for ethics—for showing us what is wise or not wise to do in our relationship to one another and to the rest of nature.

We have seen historically how we strayed from this path. Since the time of those early philosophers we have come to see our own consciousness as an ego or "I"—to set ourselves apart from nature as an objective "eye," viewing it, making theories about it, evolving a religious-scientific mechanical worldview to explain it and a mechanical technology to help us view it through lenses and exploit it through our machinery. Yet for all its early religious roots and in spite of its technological success, the mechanical worldview leaves us facing the enormous problems of today without any ethical guidance for solving them.

We have even thought we did not want such guidance, having become weary and leery of the mere mention of ethics. History has shown that ethics—as traditionally defined and dictated by religious authorities—has been used to make people obedient servants of those in power more often than it has been used for their own good. Still, when we are not handed our ethics, as was the Grand Inquisitor's flock, we are more than ever in need of guidance to overcome the anxiety that comes with freedom. We can hardly expect ourselves to take the responsibility for our actions until we have some way of judging what actions are right or wrong, good or bad.

The word "religion" comes from the Latin *re-ligio*, which means "to reconnect." Religion is our way of reconnecting ourselves to our origins, whether these origins are imagined as the creative acts of deities or seen in natural evolution, and to find some kind of guidance in these connections. Nature has very likely built into us this need to reconnect because our survival depends on maintaining connections with our origins, on understanding our relation to our co-evolving environment or ecosystem so that we may play a balanced role within it. Having given us free choice, nature also gave us the opportunity to find natural guidance for making our choices.

We have already discussed the ethical guidance to be found in the organization of our own bodies, even in the evolution of our cells. But what of the popular impression that nothing could be more *unethical* than nature? To conclude that nature is cruel and insensitive, we have only to think of a panther attacking and killing a baby gazelle, an owl pouncing on a mouse, a praying mantis biting off her own mate's head, a wasp laying her eggs inside a live caterpillar, which her children will eat from the inside out. Many species fight back viciously—plants with deadly poisons, spiky thorns, glassy stingers; a butterfly eating poison plants that would kill any bird trying to eat the butterfly; a sea urchin leaving anyone who steps on it in pain for weeks; a lion tearing up its hunter as well as its prey. Is all this not wanton, unethical cruelty?

Such arguments cannot be dismissed lightly, but let us remember that we make them from a human point of view. Indeed they reflect healthy human sensitivities which we shall discuss shortly, but to understand the Gaian system we must see it also from the perspective in which life is our planet's rocky crust rearranging itself into a multitude of species belonging to a single whole in which parts are necessarily recycled.

Right from the start, the first living packets of enzyme-driven giant molecules could build themselves only by taking in and using smaller molecules that had dissolved from rock. Many of these molecules later were built back into rock; others became parts of new bacteria. Without such recycling, where would Gaia be? The world would have filled with bacteria that would simply have died when supplies ran out, and that would have been the end of it. Earth would have ended up as lifeless as Venus and Mars are today.

Recycling is the secret of life's endless creativity, though we humans are just beginning to understand the problems our own species has caused by using things up without recycling. As supplies of "natural" molecules ran out, living holons were forced to use the "manufactured" molecules of other living holons. As we saw, for example, bacteria began consuming one another. Much later, animals had to evolve the equipment for chasing after their food rather than sitting in one spot like plants, which could make food from sunlight and local chemicals recycled by other species.

Thus large living holons evolved and maintained themselves by incorporating smaller holons and thus recycling them. What disturbs us is the fact that their food is often alive and must be killed to be eaten. But why are we so disturbed by something it is our own nature to do? Do we not protect ourselves against attack with the best weapons we know how to make? Are we not hunters and killers ourselves? Even vegetarians tear plants limb from limb, boil them to death, or crunch them up raw with grinding teeth. We cannot get along without feeding on other living things.

Is it the infliction of unnecessary pain and suffering that bothers us? Then let us note that our modern means of producing poultry and livestock, not to mention our use of research animals, is more cruel by our own standards than other species' means of killing. Our methods involve the lifelong torture of being confined in extremely limiting boxes or cages, while animals chase and kill their free prey quickly. And there is good evidence that bodies which have evolved the capacity to feel pain as a trouble signal know when they can no longer protect themselves and turn off their pain system so they will not suffer needlessly. This seems to be so for the wounded mouse in the claws of a cat as much as for the human soldier wounded in a battle with his fellow man for far less justifiable reasons.

Plants use recycled supplies in ways that don't bother us. Though the molecules they take in through their roots may have been part of large creatures, they are taken in only when the creatures have been decayed by busy bacteria. The way bacteria eat doesn't bother us either, as they and their habits are quite invisible to us.

○ ○ ○

Just as we must renew our own cells, Gaia—who is as nearly immortal as we could wish to be—cannot stay healthy without replacing her "cell" creatures. Yet natural ethics "decrees" that every holon within every holarchy must work to preserve its own life as long as possible, whether it is an individual cell, a creature, or a whole species or ecosystem.

This is the key to natural ethics—that the self-interest of *every* level or layer in a holarchy is the best possible strategy, for only by means of that strategy can mutual consistency work itself out among all levels. We've already seen how this works in our bodies, but let's look at the principle itself more closely. 'Mutual' means shared, and 'consistency' is harmony or agreement—and shared harmony is what we have called ecological balance. When every holon in the holarchy of an ecological system looks out for itself, a shared balance, or mutual consistency, results. A species holon, for instance, keeps itself healthy by producing a variety of offspring in competitive numbers, of which the healthiest are most likely to survive and become the parents of its next generation. But the species holon needs help in this selection—help from the larger holon, which is its ecosystem and which is made of other species.

Among animals, it is clear that hunters are most likely to catch the weakest members of a prey species. A hunting species thus actually helps its prey species to stay healthy by weeding out its weakest members, while the prey species helps the predators by keeping them fed. Mutual consistency often involves mutual benefit.

Most species live on a very limited diet; many will eat only a single kind of food—koalas, for example, eat only eucalyptus leaves and anteaters consume only ants. If a species eats too much of its food species, it will lose its own members to starvation. This gives the food species a chance to recover, and when it does, more of the eating species will live again. In this way species rebalance their imbalances and restore one another's health. Mutual consistency implies a dynamic—ever imperfect and ever changing—balancing process.

We saw other examples of mutual consistency between individuals and their species in the territoriality and social structures discussed earlier. Individuals fighting for their own territory, for example, do so in such a way that the whole species benefits from protection against overcrowding and inadequate resources.

Nature tests the evolving patterns of species and their ecosystems against one another to see that they are in—or are able to restore—balance, that they share harmony, or that they are mutually consistent. Whatever proves unable to gain consistency with all else around it cannot survive.

Can any of us think of a better way for life as a whole to keep itself alive and in good health? The system is worked out so that every part looks out for itself without taking more than it needs and *in doing so* contributes to the welfare of the whole! Every part thus finds its dynamic balance with every other part, working out mutual consistency in such a way that the whole Gaian system works as a single healthy organism—every part, that is, except the experimental new human species, which does take more than it needs, wantonly destroying whole other species and ecosystems in the process and killing and starving large numbers of its own species, all the while accusing the rest of nature of cruelty.

What shall we make of this human sensitivity that lives in us side by side with our own cruelty? What of this pity that we profess to feel for other creatures as we torture and slaughter them? Both aspects of this human portrait are unique; no other species demonstrates either the wanton slaughter of, or the pity for, other species. Could these feelings of pity be made to serve some useful purpose in the dance of life, as most things surviving in nature seem to do? Could the feelings serve to awaken us to our own reckless cruelty and push us toward mutual consistency with other species?

Concern for others, as we saw, evolved in mammals along with the beginnings of behavioral choice, the ability to learn, and the birth of live young that needed care and teaching. Feelings seem to be ways of directing or guiding behavior when there is choice. As we humans have the freest behavior, the greatest choice, we might expect ourselves to have the strongest and most varied feelings. And so it seems to be. A great deal of human behavior is guided by feelings—for better as well as for worse. When our feelings take over completely, we lose our ability to think about the possibilities and consequences of choice. On the other hand, when we use only our ability to think, we become cold and calculating in a way we think of as mechanical, or "inhuman." Like everything else in nature, human thought and feeling are ever in need of balancing.

But let's get back to the particular human feeling of horror at nature's cruelty. When Darwin announced that competition over inadequate resources was the sole driving force of evolution, his theory, as we saw, was quickly used to excuse cruelty in the human world—if all nature was a bloody battle "red in tooth and claw," then why should humans be an exception? The competitive exploitation of resources and labor by the rich as they built an industrial world was thus justified on the grounds that it was natural!

Even now there are sociobiologists who believe that aggression is our innate and therefore unchangeable animal heritage, though they make no claim that it is ethical. The anthropologists who study various human cultures have told us that ethics is really no more than a set of behavioral codes specific to individual ethnic groups. What is ethical in one culture is unethical in another, so there is no point trying to find a common human ethics.

Yet a common human ethics is what we now need more than anything else—an ethics to guide us in our behavior toward one another and toward the natural world to which we belong. Our basis for such an ethics is very different now that we no longer see nature as just a bloody battleground for competitive struggles over limited resources. Competition is merely one aspect of nature's creative organization into mutually consistent holons within holarchies. What we see clearly is that every holon's health depends on the health of the larger holons in which it is embedded. Thus every holon, in looking out for itself, must also cooperate with other holons to help look out for their larger holon's interests.

This, as we said, is the heart of ecological ethics—the self-interest of any particular holon, whether a cell, a body, a society, a species, an ecosystem, or a whole living planet, balanced in the mutual consistency of the whole and all its parts. For us this means recognizing how much we affect the living planet of which we are part and on which our continued existence depends. To look out for our own interests *requires* that we know the interests of our whole environment, which means our whole living planet. Our free choices, in order to serve our own long-range interests, must serve those of other species as well, for natural ethical behavior is that which contributes to the health of the whole Gaian system.

Our history has brought us to the shortsighted adolescent selfishness of warfare, hatred, distrust, and reckless destruction of our own environment. We have long-standing habits of believing that all nature is human property, and so we take land and resources from one another for reasons of profit. It is high time for us to realize that maximizing individual profits minimizes human social stability and welfare, while maximizing common profits destroys our natural life-support system. If we want to survive as a species we must learn to change our ideas and our lifestyles to live in a balanced recycling economy like the rest of nature.

In fact, it is high time to realize that all our old habits and

ECOLOGICAL ETHICS

vested interests, even if they form our individual and national identity, must be fundamentally changed. The changes required are deeper and more far-reaching than any revolutionary leader has ever demanded or even dreamed of demanding. And yet we can make those changes peacefully, and everyone can win.

The problem is that they cannot be made at the point of a gun. They must be made voluntarily and that is perhaps more difficult. The profit motive is so ingrained in Western society, for example, that scientists have actually criticized nature on the grounds of unprofitable inefficiency, pointing out that photosynthesizing plants use only a small fraction of the energy available in sunlight. Can such people learn to appreciate the fact that plants extract exactly as much energy as they need for themselves and to keep their environment's careful balance of energy exchange?

The cure for the anxiety of freedom is the security of having some way of knowing what to do. If we agree to consider ethical human behavior whatever we sincerely believe, on our best knowledge, to be healthy for our own species and healthy or at least harmless for other species, for the environment, for our planet, then we have such a guide.

Our age-old re-ligious quest for reconnection with origins has been the search not only for our origins, but for our creator as an inspirational source of guidance and security that would lead us to a better life. In the childhood of human civilization we imaged this source as parental deities cast in human images. Then, in our adolescent cheek, we rejected them all, believing there was nothing greater or more intelligent in all the universe than ourselves. Now, on the brink of maturity, we can see that our early intuitions were valid. The source of our creation is indeed an inspirational being far greater and wiser than ourselves; a being that has nurtured us and can guide us to a better way of life—not a perfect superhuman parent, but the imperfect yet wonderfully resourceful planet of which we are one part, and which may itself be a part of a much greater being. Have we the maturity to heed it?

19

Cosmic Continuation

Life, as we have understood it in this book, is the self-organizing, or autopoietic, activity of matter, driven by energetic interactions between great and minute events, between the tendencies of matter/energy to differentiate and to integrate. In this view of life, matter/energy arranges itself into bounded but interacting living systems, perhaps on a galactic or supergalactic scale, certainly on a scale as large as our entire planet, as well as in the microscopic world of bacteria. From our present perspective and limited knowledge, we can see that planetary life has evolved the most active and complex systems.

In the older, still prevailing mechanical worldview, life is understood as an incidental part of the universe rather than as its essential tendency. In this view, the lifeless universal mechanism had already been grinding along for more than ten billion years when some of the nonliving matter on rare planets—at least one—was accidentally converted into living matter. This was the only way life scientists could fit their discoveries into established cosmic models.

Because modern science was founded by astronomer-physicists, their mechanical mathematical models of the cosmos were accepted as the basis of all science. Physics is still considered the most basic science—the one responsible for explaining nonliving matter, the one responsible for explaining how the cosmos is formed. We can scarcely guess how far the organic worldview might have been developed by now if biologists instead of physicists had played the leading role in science—if physicists had had to fit their discoveries into the model of an organic, live universe.

Biologists—who work with living organisms and see them reproduce, develop as embryos, care for themselves—have often found it hard to fit life into the mechanical worldview. But biology simply did not have the status of physics, and so biologists had a hard time challenging the physicists' models of nature. Now that we can see Earthlife as part of a self-creating galaxy, as planetary crust transforming itself into a web of creatures and environments, it is clear that living things are not built up from pieces and shaped from the outside, as is machinery. Living things are not assembled from molecules here and there on some planets and then in turn assembled into ecological systems. Rather, some whole planets develop the metabolism of living beings, coming ever more alive in the great flow of energy between their stars and themselves, gradually packaging their crust material into ever more species that weave their own changing environments.

As yet we don't know what role living planets play in their galaxies and within the larger systems of our universe. Our galaxy has differentiated itself from a protogalactic cloud into a complex system of nucleus, stars, star systems, clouds, planets, comets, and other parts. Most likely, only a few of many planets—as with the spores, seeds, and eggs of Earthlife—come to develop. Perhaps those that do will integrate themselves like bacteria into nucleated cells, or nucleated cells into multicelled organisms, or humans into a body of humanity—if they are not already part of a larger being.

There are endless mysteries to solve. A great variety of organic molecules, for instance, come through interstellar space, apparently from the center of our galaxy. Yet we don't know how they can be formed there or even what this center is made of, because it is hidden from us by great clouds which our telescopes cannot penetrate. Nor have our telescopes yet revealed to us the planets belonging to any star other than our sun, though we know they must be out there. Our present understanding may be something like that of mitochondria in one of our own living cells trying to understand the organ "galaxy" and "universe" body they are part of without being able to see or even guess very much of what these larger holons are all about.

Nevertheless, the organic worldview will give us ever more perspective on the limits of mechanics in explaining the universe, and it will also give us ever more power to go beyond those limits, just as non-Euclidean geometries and the theory of relativity broke through the limits of Euclidean geometry and classical physics.

Science does on occasion reach out boldly into the new, but on the whole it tends to be conservative. Most physicists are still reluctant to shift their perspective, even though they are having ever more trouble fitting their new discoveries into their older mechanical models. To their surprise, they have found that the most basic properties of matter—properties and limits matter had from the very beginning of the universe—are precisely those *necessary* for life to evolve. In fact, they make life inevitable. This, of course, is completely inconsistent with the idea that life is accidental and incidental to the universe. But instead of changing their cosmic model from a mechanical one to an organic one in which the self-organization of living systems is natural to cosmic matter, some physicists are going back to the old view of purposive mechanics. They propose an *anthropic principle,* which says that the fine-tuned properties of universal matter are best explained by assuming its goal from the start was to produce man! As physical matter by itself can have no purpose or goal, the anthropic principle is interpreted as implying a God-mind Creator.

There is, of course, no way to prove or disprove the existence of such a God-mind. As we said earlier, our worldviews are *our* creations, based on limited information about reality, and we have no information at all on such matters. The existence of a Creator has always been for us humans a matter of faith—a belief that a Creator must exist if no other explanation of creation can be found.

In the past century, scientists have worked at explanations that are testable as much as possible by experiment and that are based on the fewest possible assumptions. This is true in particular of those explanations that do not require the assumption of a Creator. Yet the question of how life began in a mechanical universe without a Creator, and how it keeps itself going, has not been convincingly answered within the scientific mechanical worldview. A century ago, mechanical explanations of life as an accidental assemblage of molecules, changing form by further accidents and selected into different species by mechanical pressures of competition, were convincing enough, but they have become ever more inadequate as our knowledge has grown. Now we must decide whether to go back to a model of creation brought about by a purposive deity or work further on the model of a self-organizing universe.

Not long after the mechanical theory of evolution was proposed by scientists, it was opposed by some philosophers, including Henri Bergson and Pierre Teilhard de Chardin, and Alfred

North Whitehead, who worked out organic models or theories in which life is seen as inherent in the universe. Bergson opposed the idea of purpose in nature, but proposed the existence of a mysterious life force that is separate from and struggling with matter in an attempt to organize it. Scientists tended to reject his model for this reason. Teilhard de Chardin, though he explained life as the natural evolution of self-organizing matter, saw evolution as purposive, leading by way of mankind to a "God-Omega point," and his work was also rejected by most scientists. Whitehead was too philosophical and obscure for practical scientists, but his talk of organics and God would have put them off in any case.

We have also seen that at least a few scientists worked at theories of organic creation within matter itself. One such scientist was Charles Darwin's younger Russian contemporary, Vladimir Vernadsky, whose idea of life as a metabolically active "disperse of rock" has pervaded this book. Another was George Hutchinson, the Yale University professor who spread Vernadsky's ideas in America. James Lovelock and Lynn Margulis acknowledge geologist-physician James Hutton's concept of a living earth as a forerunner to the Gaia hypothesis. Erich Jantsch's "self-organizing universe," which also influenced the writing of this book, shows the interplay of the largest and smallest events of cosmic matter/energy producing the ever more complex systems of an essentially live universe. None of these scientific models include the notions of purpose or life force or God.

The mystery of life begins clearing up as soon as we stop thinking of the cosmos as a mechanical assembly of atomic, astral, and galactic parts—as soon as we begin to see that cosmic "parts" form themselves from, and are formed within, larger wholes, rather as species form themselves within, and are formed by, ecosystems in their evolution. Is it not likely that life "here below" behaves essentially the same way as things do "above" in the greater cosmos, as the ancients claimed?

In Chapter 2 we saw that forms as simple as whirlpools, in water or in protogalactic clouds, take in matter, maintain their form in shaping its flow, and give off used matter to their environment—the basic pattern of any living thing's organization in relation to its surround. Though such whirling clouds do not transform materials as required for autopoiesis, they certainly do when they evolve into full-fledged galaxies, showing that the process from pre-life to life is continuous. Whether galaxies are self-bounded and exchange materials across these boundaries is still an open question; galaxies *are* known to split and to merge. In any case, living holons gain their identity and relative autonomy by organizing their own form and function, yet they remain dependent on their environment or larger holon to supply their resources and absorb their wastes.

Life is also less mysterious when we see how it organizes itself through interactions between vast galactic whirlpools and tiny energetic atoms, between the energy of our sun star and the molecules of our planet's crust, between moon-pulled tides and the tiny liposomes they bathe and dry out in cycles. In many ways, the evolution of a self-organizing living planet is a less mysterious, more elegant, and more satisfying conception of life than the idea that nature's awesomely intricate living patterns and harmonies should have been brought about either by the mechanical accident and assembly of parts, or by the purposive design of an invisible mind.

Still, scientists may continue for some time to vacillate between ideas of accident and purpose to explain how non-life becomes life. It wasn't easy to see matter and energy, wave and particle, time and space, as aspects of the same underlying unity; it may be even more difficult to see life and non-life as aspects of the cosmic process or as relative organizational states of cosmic matter. We have no problem seeing the process of life in a puffball bursting to scatter its spores, but we do tend to have difficulty seeing the process of life in a star bursting to scatter its dust. Though we ourselves are reorganizations of the same dust, we have been told by astronomer-physicists that stars belong to the domain of non-life.

The further development of an organic model of the universe clearly requires the cooperative efforts of biologists and physicists as well as other scientists. With so much to learn about nonmechanical autopoiesis or self-creation, this is a most exciting and challenging time of discovery and new understanding. Most important among the developing efforts is the promise of seeing ever more clearly just how we humans fit into the great cosmic holarchy of life and how we may learn to cooperate in promoting its health along with our own.

○ ○ ○

As we have seen, our historic worldviews—our images of who and what we are in relation to all nature—have been, on the one hand, limited by narrow perspective and, on the other, unbounded in egotism. For thousands of years we considered ourselves God's favorite creatures; then, when we had no more use for God, we saw ourselves as the pinnacle of natural evolution. In both views nature was ours to command and exploit as we liked. Only when we ourselves began suffering from the damage we had done to our environment did we begin to gain a more realistic view of ourselves as one species among still uncounted others on whom we depend. Shall we now abandon our newfound grain of humility for an anthropic principle in which not only the earth but all the universe is regarded as having been created for man?

In truth, ours is a middle-sized planet circling an ordinary star at the edge of a common type of galaxy; our world is now in its middle age, about halfway through its expected life. Its oldest living creatures are bacteria, the first species with which it came alive billions of years ago and which are still the basis of its self-regulation systems and of all other life forms that have evolved since. Its newest living species are mammals that evolved only millions of years ago, yet not a single species of early mammals survives today. The coming and going of new species began to speed up with the self-organization of nucleated cells and occurred even more rapidly after the invention of multicelled creatures. In fact, all such forms "beyond" bacteria can be seen as their inventions, vehicles for their continuing survival.

There is little reason to think that humans will change the pattern of species flux and survive for the rest of the life of our planet without further evolving. We haven't even the patience to wait for natural evolutionary changes, but chafe at the bit to redesign our own DNA as well as to create mechanical humanoids. In light of this, is it not preposterous to propose ourselves as the goal of all creation? Would it not be more appropriate and wise to recognize ourselves as a bold trial species in the Gaian life system? As a new kind of creature that may not survive very long at all if it doesn't learn to play a humbler and healthier role than it has thus far?

On our planet, we are indeed unique for the range of our conscious free choice of behavior, the range of our technological prowess, and our ability to foresee, plan, and act for the future. But the very evolution of such abilities on our own planet suggests they may well have evolved on many other planets around the universe. The same kind of matter exists everywhere and seems to undergo the same processes all over the cosmos. There is no good reason not to assume that planets come alive and evolve in essentially the same way everywhere. This assumption has been made widely—at least it *was* until the anthropic principle of human uniqueness was devised—and underlies our ongoing efforts to reach and communicate with intelligent life in other star systems.

In these communication efforts, scientists assume that intelligent species on other planets will have invented a mechanical technology composed of mathematics and machinery essentially similar to our own. This assumption is made in the belief that intelligence—the gathering and using of information to gain understanding—can grow to human proportions only through our kind of technology, through a similar co-evolution of conscious thought with the productive use of manipulative organs such as our hands. This idea rules out the possibility that creatures such as cetaceans could be as intelligent as humans, and it would be wise to feel less sure of ourselves on this score until we know more about the whales and dolphins.

It is interesting to consider that several of our own technological inventions—including sonar, diving equipment, insulated clothing, and high-speed long-range communications systems—all evolved within cetacean bodies, which function comfortably at widely varying pressures and temperatures as they move freely about the seas covering most of the earth's surface. We have no present way of knowing how far such naturally evolved "internal biotechnology" can go or how it can be used by highly evolved brains—whether it could contribute to the formation or function of living systems beyond our planetary Gaia.

We ourselves, in addition to our external mechanical technology, have an external organic technology that began with our ancient use of natural herbs as medicines, the deliberate alteration of plants and animals by selective breeding, the creation of artificial agricultural ecosystems, the use of natural predators to control pest species, the use of bacteria and yeasts in making bread, beverages, and yogurt, and so on. All of these technologies were already known in the Stone Age when humans lived peacefully, worshiping the power to give life rather than the power to take it.

Though modern science has focused its attention primarily on inorganic mathematics and machinery, organic technology—which in many cases is really the discovery and use of the internal technology long ago invented by bacteria—is suddenly developing rapidly. From breeding microbes both as healing antibiotics and as germ-warfare weapons, we have progressed to replacing our damaged tissues and organs with those of other humans or animals. We use bacteria to clean crude oil and to manufacture biodegradable plastics; we cut and splice their DNA programs to make them manufacture particular substances for us. We produce and patent new plants and animals and create ever new kinds of artificial ecosystems.

This rapid progress in biotechnology will likely get us into as much trouble as our nuclear-age mechanical technology if we don't make equal progress in understanding life systems and their dynamic ecological balance. Only if it is used with understanding of and respect for living systems can biotechnology offer the possibility of working *with* life *for* life.

We can imagine, then, at least three kinds of technology that intelligent beings of other planets might have in various combinations: mechanical technology, internal biotechnology, and external biotechnology. So far, our space communications systems have been designed only to make contact with beings that have mechanical technology. Our efforts to communicate with the cetaceans of our own planet, which might be the best way to practice for interstellar communications, are very meager.

We do not know whether galactic life systems larger than those confined to single planets already exist or whether they are yet to be formed, but in our imagination, in our science fiction—much of which has already become reality—we rehearse the possibilities. A galactic "Gaian" system would require communications if not space travel. If live planets are the cells of such a galactic system, their living creatures may not have to come into physical contact with one another, any more than the mitochondria of our liver cells need to meet the mitochondria of our bone cells. A common information-communications system may unite them effectively. At present, time is as much a barrier to interstellar messages as to physical space travel, but our concepts of time are still in flux, and solutions may yet present themselves.

○ ○ ○

Before we consider from our own point of view how we might become part of an interplanetary or intergalactic life system, shouldn't we consider how we might look to other members of such a system? We have seen the Gaian system and ourselves from various perspectives within or just outside it. Let us now try looking at ourselves from the point of view of some intelligent species that can observe us from another star system.

If they have detected us with radiotelescopes, or other means of reading radio signals, the earth, to them, would at first look like a larger companion star to our sun. On closer look, however, the earth's tremendous chemical activity would alert them to the presence of a live planet. Thus they could interpret the radio signals as those of an intelligent species.

If they could focus on us humans and learn what we are and what we are up to, would they consider us an intelligent form of life?

Surely it would strike them as most peculiar that we destroy the environment on which we depend. No intelligent species would knowingly pollute its air, water, and soil to the point of endangering itself. It would hardly cram itself into communities of concrete that sealed the species off from natural processes and made its air unbreathable with its own wastes when there was plenty of space on the planet. They would wonder why we destroy the natural ecosystems of our planet to grow our own food without preserving variety and recycling water and nutrients to prevent the land from turning into deserts. They would note that we deliberately overload our planet's delicately balanced atmosphere and thus overheat the planet itself with carbon dioxide as we hack down the tropical forests and destroy the coastal areas that might reestablish the balance. They would wonder how we could know that our whole planet's life system is driven effectively by solar energy and not use this safe energy source extensively ourselves.

Could they consider us intelligent after they see that we are quite capable of providing a comfortable life for all humans, yet choose to devote enormous resources to escalating an arms race rather than spending them on the welfare of our people? Would an intelligent species overproduce food for some of its members while millions of its young die annually of starvation because others are deprived of the means to feed themselves? Would our observers not wonder that the males of our species had long ago declared the female half of the species inferior, largely excluding them from their former positions of authority and management, at the same time often shifting social priorities from life-giving to life-taking?

They would see us systematically exterminating other species and whole ecosystems, even the other large-brained creatures most like ourselves, such as dolphins and whales, elephants, chimpanzees, and gorillas, though they are peaceful and inoffensive to humans. They would see our leaders, many of whom hold their positions by popular consent, maintaining hostilities that threaten nuclear warfare, which would destroy all parties and create a planetary nuclear winter. What would they think of our popular Star Wars cult and of our use of that name for a weapons system?

Our scientists think it inevitable that any extraterrestrial species with a technology essentially similar to ours would discover nuclear energy. If it is, or was, a warring species at the time it made this discovery, it would have faced a crisis similar to our own—a crisis of choice between species suicide or species cooperation. Any species that has survived the discovery of nuclear power, that is, can be expected to be peaceable. This leaves our expectations of alien belligerence unrealistic and perhaps—if it comes to a meeting—embarrassing. Those who might have wanted to engage in star wars have likely already blown themselves up. We might also expect that any intelligent species becoming aware that its technology was destroying its own ecosystems would change its ways promptly rather than endanger its own survival.

Are we a cosmic anomaly—a species once mutually consistent with its ecosystem that became rather suddenly, over a period of a few thousand years, hostile and destructive to itself and its living planet? May we attribute this peculiarity to the temporary condition of youth—to unprecedented freedom during the heady stage of species adolescence? Have our observers reason to hope we shall soon grow up and become wiser? Let's consider the evidence in our favor.

We have begun to understand and be concerned with ecological balance; we are beginning in at least some areas to protect endangered species, reduce our pollution, and give the rest of nature a chance to clean up and restore its balance. Unfortunately these constructive efforts are very far from balancing our continuing destruction. Still, there is hope that we may increase these efforts dramatically when we finally recognize ourselves as part of a living being whose delicate balance is tailored to our needs, but which may eliminate us if we force it to reorganize itself to cope with our reckless destruction of that balance.

The threat of all-out nuclear war is actually lessening as we take to heart our scientists' predictions of its dire consequences for all humanity and work more seriously toward real disarmament. This suggests that we *do* respond to our knowledge by changing our ways, though disarmament efforts so far have been minimal.

Our technology is well on the way out of the industrial age of heavy steel and polluting fossil fuels and into an information age of lightweight transistor technology and benign energy sources such as sunlight, wind, waves, hydrogen, and alcohol. If we put our future welfare ahead of immediate profit motives, this transition could be completed rapidly.

Agricultural research is beginning to look more seriously at older, even ancient, methods of natural pest control and crop rotation and variation, recognizing that monocultures sustained by chemicals are rapidly destroying the land and polluting the waterways.

Our worldwide economic system, our transportation and communications technology, our information revolution, have bound us into a body of humanity that is now being pushed for the sake of its survival to evolve from competition to cooperation among nations and with our environment. We can see that the integrative evolutionary forces which produced protists and multicelled creatures are now pushing us to complete our own organizational task. We can see that we only prolong and aggravate our biggest problems by resisting this evolution with habitual fears and hostile competition inspired by profit motives. We can use our gift of freedom to make up for our lack of innate limits to territorial and aggressive behavior by channeling these into constructive negotiation and sharing.

We can recognize that the strength and resilience of living systems lies in their diversity, and stop trying to make ourselves all alike. We can strengthen and support efforts such as the United Nations, which are as critical to the organization of the body of humanity as were nuclei to eukaryotes and brains to animals.

Perhaps most important of all, we can awaken ourselves to sexual inequality not merely as an injustice to women but as a deplorable waste of half our species' talents for creative management and nurture of the species as a whole. As long as women remain a tiny minority in positions of human leadership, they will be pressed on the whole to conform to the established male model of society based on mechanical organization, bureaucracy, competition, conquest, and profit. Only when women assume equal leadership will they be free to express effectively their abilities in organic organization, networking, cooperation, nurture, and mutual benefit.

Certainly we would be foolish to continue our environmental destruction and our hostilities to the point of species suicide when we are surrounded by clues to exciting, creative, natural solutions for our greatest problems. If we accept ourselves as an adolescent species in crisis and face the challenge of taking on mature responsibility for our freedom with courage and enthusiasm, we stand an excellent chance of growing up and reaping the benefits of maturity.

◯ ◯ ◯

How rewarding it would be if, when we had our first communications with extraterrestrials, we could report having solved our great problems of inequality, hunger, pollution, devastation of ecosystems, and nuclear threat and also report that we are ready to face new challenges with appropriate pride in our species. Perhaps we will be called on to protect Gaia by detecting and diverting massive meteors or planetoids; perhaps if we come to understand and help foster Gaia's healthy ecosystems, we will be able to bring life to Mars or some moons of our solar system, thus spreading Gaia's seed.

Whatever our dreams of such roles, or of cooperating in a larger cosmic life system, our immediate task is here at home. On the whole, there seems to be good reason to believe our recklessly egotistical and destructive phase is coming to an end with new knowledge that leads us back to ancient wisdom. For the first time in our history we can look back on that history and see that our self-image as naturally belligerent creatures is belied by tens of thousands of years of peaceful partnership societies. We are capable of regaining our reverence for life, of replacing the drive to conquer with the will to cooperate. For the first time in our history we can actually see our whole planet and recognize it as a living being—and we can understand that we are not its privileged rulers, as we thought for so long, but only one part, and not even an indispensable part, of its body.

The more we learn about nature, including human nature, the less sense it makes to create deities in our own image and tell ourselves myths about being their favored creatures. The age-old intuitive feeling that there *is* an intelligence greater and wiser than ourselves and to which we belong is now validated by science, and we can now see that our live parent planet and our whole living cosmos are far more beautiful and awesome in the reality of their self-creation than is any myth we made as we struggled to develop our knowledge. At last our scientific and religious quests can merge in the recognition that self-creating nature—Gaian and cosmic—is our physical source, the wellspring of our ancient inspiration, and the experienced guide we have always sought, the guide we need more than ever now that we stand on the brink of maturity.

Epilogue

Books are not published overnight, and so there is time for the world—and the author's thoughts about it—to go on evolving between the time a book is completed and the time it actually goes to print. As I apply my own ideas, as expressed in this book, to the world around me during this time, it seems to me that humanity's adolescent crisis is still deepening. The wanton destruction of nature and the production of nuclear weapons continue despite our widespread knowledge that what we are doing threatens our own and our children's existence, the whole future of our species.

There is an eerie feeling of science fiction in the daily media reports. It is as though our journalists are reporting a species' suicide objectively, as though it is not their own that they observe. Along with all the sensational news on ozone holes with their threat of increased radiation and the now tangible greenhouse warming attributed to the production of certain gases and the razing of irreplaceable tropical forests, we see daily reports of how pesticides, fertilizers, monocultures, and even irrigation schemes are destroying the agricultural land that feeds us.

The World Bank itself has admitted that its development schemes are an ecological disaster, that they spend enormous sums of money to create deserts in the name of making gardens. Yet we go on spraying chemicals into the air, cutting forests, "developing" the land into technologically produced deserts.

Patents on biotechnologically produced microbes, plants, and animals are already a new source of international rivalries and conflicts, while their consequences for ourselves and our environment are no more predictable than were the deserts created by high-tech agriculture.

Why can't we predict the consequences of our own actions? Because we don't know how Earthlife "works." It remains true that we are no more competent to run our planet than we are to run our own livers, and we don't even understand why it is not wise to try. The belief that we humans can manage this planet by refining our technology and inventing more of it has not abated despite the constant news of technological damage done. I am far from opposed to technology, but the kind, the manner, and the extent of its use is of critical import.

Unfortunately, scientists continue to resist the Gaian conception of the earth as a living being of which we are part and which knows how to take care of itself far better than we do. Science remains divided into categories such as biology and geology, and their respective practitioners are equally divided in their views on Gaia theory. Debate about Gaia theory has come to resemble a court trial whose purpose is to decide whether the living things *on* planet Earth, defended by biologists, should get credit for controlling its weather, temperature, and chemical conditions or whether the nonliving environment, defended by geologists, could manage such matters without the assistance of life and even, as is sometimes implied, without the interference of Gaia.

Control and management, not life, seem to be the issues at stake. The version of Gaia theory scientists will most likely come to agree on is one in which living things and nonliving things are seen as built into an engineered system of mutual-control cybernetics. For if our planet resembles one of our cybernetic mechanisms, there is still hope we can learn to control it with our technological expertise.

Someday—if our species survives, as I continue to hope it will—I believe history will record that the recognition of Gaia as an autopoietic system, a living planet, an earth organism, was resisted in the twentieth century for much the same reasons that Copernicus's heliocentric theory was resisted in the sixteenth and seventeenth centuries: because it violated strong established beliefs and threatened vested interests. During the Renaissance, men did not want to give up the idea that they were the best and most loved of God's creatures and that God had therefore put them in charge of the most important piece of property in the whole universe—the planet at its very center. Now they cannot accept that they are not the pinnacle of evolution, the "brains" of our planet, the species fit to rule all else—even that they are the "anthropic" goal of the universe itself.

The concept of Gaia as a living—autopoietic—being is actually not a theory to be tested. If we agree that autopoiesis is an adequate definition of life, and if we agree by evidence such as that presented in this book that the earth meets this definition, then we have agreed the earth is a living being. And then we are free to make Gaian theories about its specific functions as we make theories about our own physiological functions.

Until we accept this concept and make the profound transition in worldview it implies—the transition from seeing ourselves as actually or potentially competent in managing Gaian mechanisms through our own technology to recognizing ourselves as a new and still highly experimental organ evolving within a larger organism—I do not believe our practice will change fast enough to save us, to prove the experiment successful.

We are not held up by lack of awareness of ecological damage, for awareness has spread and increased almost like wildfire. But despite the damning reports of the Worldwatch Institute, the increasing admissions of mismanaged development such as that of the World Bank, the emotional impact of Greenpeace; despite the extensive work of the International Union for the Conservation of Nature (IUCN), the United Nations Environment Program (UNEP), and other organizations planning sustainable development, our juggernautlike demolition of nature continues. Despite awareness of the possibility of a nuclear winter and other consequences of nuclear war, despite small U.S.-USSR gains in disarmament, nuclear weapons continue to be produced and stockpiled faster than they are destroyed. The fact is—and almost everyone I speak with agrees and is alarmed by it—that too many men with the means still show an adolescent desperation in their willingness to poison, parch, or blow up parts of the earth for the sake of short-term profits, refusing to take the long-range view or the responsibility for a long-term future.

Some pre-publication reviewers of this book felt the value of its scientific arguments was diminished by its feminist emphasis. This puts me in a bind, for I believe resistance to accepting ourselves as part of a live planet is profoundly linked to the loss our species has suffered as a result of the failure to give women equal say in forming worldviews and managing human affairs.

Scientists, for better or worse, are the priests of our society; when they say the earth is alive, we shall all believe it. But science, as Brian Easlea has documented in his book *Fathering the Unthinkable,* is built on an intensely male perspective and drive to conquer and rule nature. It is this perspective and drive that cries out to be modified by a female perspective, but though many women have now been trained as scientists, they are trained to accept an established worldview and to perform competitively within it, as are men. As Jim Lovelock pointed out in his preface to this book, few scientists have attained as much independence as he and I have.

One critic suggested that my fiery feminism, carried to its logical conclusion in terms of my body analogy, "would demand that half of bone cells be immediately allowed to become muscle cells, in the interest of equality." I am disappointed that this reader did not get my message of respect for the body in both diversity and organization; I would not think of making any demands on it at all, and the idea of making bone cells into muscle cells is no more appealing than my negative "what if?" illustration of the heart persuading the liver to become a heart. Rather, I suggested *we* follow the body's lead. Body cells are not divided overall into two kinds, such as male and female, so no direct analogy is appropriate. What we do see is that the coordinating nervous system of a healthy body "listens" equally to all kinds of cells and serves them without discrimination. In the body of humanity, as well as within each of its nations, I see no way this could happen except through the proportional representation in all governments of men and women, and of the various ethnic groups in their populations.

Not only government but all aspects of human history undervalue and underrepresent women. There can be no doubt that women have always played primary sustaining roles in domestic, agricultural, and cultural affairs, maintaining homes, fields, crafts, and cultural traditions including history, religion, education, and art, but their activities, with few exceptions, were not historically valued and recorded from the time of the great invasions as were the rule and deeds of men in war and in peace. The historical record of politics, economics, religion, philosophy, science, art, and literature is overwhelmingly male, and even the most enlightened cultures from ancient Athens to modern Western nations did not even give women the vote until well into the twentieth century.

There is no reason to distort the historical record to support feminist or anti-feminist positions, no reason for modern men to feel either proud or guilty about a past they had nothing to do with, and no reason for women to hold them accountable for it. What matters is what all of us do in the present to ensure a healthier future. The historical record is not cited as an attack on men but to point out, as Riane Eisler does, that thousands of years of dominance-culture habits can still be undone to all our benefit with a return to partnership in contemporary society.

How can we increase the courage of women to speak about and to act on the convictions I find universal among them—convictions I also find among many supportive men—that no conflict among humans is worth sacrificing our children to, that there is no more important priority than feeding people, that our separation from one another, our distrust and hatred, are rooted in our failure to care for one another, to love by touching, feeding, cleaning, and nursing, as women do? These undervalued concerns and activities, the respect for and love of life, are what will heal the world if they are recognized and acted upon, not just among individuals and in homes, but in our greater society as well—in economics, politics, even science. There is no other way.

Bibliography

Abraham, Ralph H. 1981. "Dynamics and Self-Organization." In F. F. Yates, *Self-Organizing Systems: The Emergence of Order*. New York: Plenum.
Augros, Robert M., and George N. Stanciu. 1986. *The New Story of Science*. New York: Bantam.
Balandin, R. K. 1982. *Vladimir Vernadsky*. Outstanding Soviet Scientists Series. Moscow: Mir Publishers.
Barnett, Lincoln. 1957. *The Universe and Dr. Einstein*. New York: Bantam.
Barrow, J. D., and F. J. Tipler. 1986. *The Anthropic Cosmological Principle*. Oxford, England: Clarendon Press.
Bateson, Gregory. 1972. *Steps to an Ecology of Mind*. New York: Ballantine.
Bateson, Gregory. 1979. *Mind and Nature: A Necessary Unity*. New York: Dutton.
Bergson, Henri. 1911; reprint, 1983. *Creative Evolution*. Lanham, Md.: University Press of America.
Berry, Thomas. 1987. "Thomas Berry: A Special Section." *Crosscurrents* 38 (2, 3).
Bohm, David. 1957. *Causality and Chance in Modern Physics*. New York: Harper Torchbooks.
Bohm, David. 1980. *Wholeness and the Implicate Order*. London: Routledge & Kegan Paul.
Boon, Jaap. 1984. "Tracing the Origin of Chemical Fossils in Microbial Mats." In Y. Cohen, R. W. Castenholz, and H. O. Halvorson. *Microbial Mats: Stomatolites*. New York: Alan R. Liss.

Boorstin, Daniel J. 1985. *The Discoverers: A History of Man's Search to Know His World and Himself.* New York: Vintage.

Briggs, John P., and F. David Peat. 1986. *Looking Glass Universe.* New York: Simon & Schuster.

Bronowski, J. 1974. *The Ascent of Man.* Boston: Little, Brown & Co.

Brown, L. R., E. C. Wolf, S. Postel, W. U. Chandler, C. Flavin, and C. Pollock. 1985. "State of the Earth," *Natural History,* April.

Calder, Nigel. 1978. *The Key to the Universe: A Report on the New Physics.* New York: Penguin.

Campbell, Joseph. 1968. *The Masks of God: Creative Mythology.* New York: Penguin.

Cannon, Walter. 1939. *The Wisdom of the Body.* New York: Norton.

Capra, Fritjof. 1975. *The Tao of Physics.* Berkeley, Calif.: Shambala.

Capra, Fritjof. 1982. *The Turning Point: Science, Society and the Rising Culture.* New York: Simon & Schuster.

Charlson, R., J. Lovelock, M. Andreas, and S. Warren. 1987. "Oceanic Phytoplankton, Atmospheric Sulfur, Cloud Albedo and Climate." *Nature* 326:655.

Chorover, Stephen. 1976. *From Genesis to Genocide.* Cambridge, Mass.: M.I.T. Press.

Darwin, Charles, and T. H. Huxley. 1983. *Darwin and Huxley: Autobiographies.* Edited by Gavin de Beer. Oxford: Oxford University Press.

Dawkins, R. 1976. *The Selfish Gene.* Oxford, England: Oxford University Press.

Dostoevski, Feodor. 1879; reprint, 1977. *The Brothers Karamazov.* New York: Penguin.

Dunhamn, Barrows. 1960. *Thinkers and Treasurers.* New York: Monthly Review.

Durant, Will. 1961. *The Story of Philosophy.* New York: Simon & Schuster.

Easlea, Brian. 1983. *Fathering the Unthinkable: Masculinity, Scientists and the Nuclear Arms Race.* London: Pluto Press.

Einstein, Albert. 1950. *Out of My Later Years.* Scranton, Pa.: Philosophical Library.

Einstein, Albert. 1961. *Relativity: The Special and General Theory.* New York: Crown.

Eisler, Riane. 1987. *The Chalice and the Blade.* San Francisco: Harper & Row.

Fleischaker, G. R. 1988. *Autopoiesis: System Logic and Origins of Life.* Ph.D. dissertation, Boston University.

Gehrz, R. D., D. C. Black, and P. M. Solomon. 1984. "The Formation of

Stellar Systems from Interstellar Molecular Clouds," *Science*, 25 May 1984.

Gimbutas, Marija. 1975. "Figurines of Old Europe (6500–3500 B.C.)." *Transactions of the International Valcamonica Symposium on Prehistoric Religions*. Capo di Ponte, Brescia, Italy: Edizione del Centro.

Gimbutas, Marija. 1980. "The Temples of Old Europe," *Archaeology*, November-December.

Gimbutas, Marija. 1981. " 'The Monstrous Venus' of Prehistory, or Goddess Creatrix." *Comparative Civilizations Review*, 7 (Fall 1981).

Gimbutas, Marija. 1982. "Megalithic Tombs of Western Europe and Their Religious Implications." *Quarterly Review of Archaeology* 6(3).

Gimbutas, Marija. 1982. *The Goddesses and Gods of Old Europe, 7000–3500 B.C.* Berkeley: University of California Press.

Goodall, Jane. 1983. *In the Shadow of Man*. Boston: Houghton Mifflin.

Gould, S. J. 1977. *Ever Since Darwin: Reflections in Natural History*. New York: Penguin.

Gould, S. J. 1984. *The Mismeasure of Man*. New York: Penguin.

Graves, Robert. 1957. *The Greek Myths*. New York: Penguin.

Gray, Michael W. 1983. "The Bacterial Ancestry of Plastids and Mitochondria" *BioScience* 33(11).

Grene, David, and Richard Lattimore, eds. 1956. *The Complete Greek Tragedies*. Chicago: University of Chicago Press.

Haldane, J.B.S. 1985. *On Being the Right Size and Other Essays*. Edited by J. M. Smith. Oxford, England: Oxford University Press.

Heisenberg, Walter. 1975. *Across the Frontiers*. New York: Harper Torchbooks.

Ho, M. W., and P. T. Saunders, eds. 1984. *Beyond Neo-Darwinism: An Introduction to the New Evolutionary Paradigm*. London: Academic Press.

Ho, M. W., P. T. Saunders, and S. W. Fox. 1986. "A New Paradigm for Evolution," *New Scientist*, 27 Feb.

Hooker, Michael, ed. 1978. *Descartes*. Baltimore: Johns Hopkins University Press.

Jantsch, E. 1980. *The Self-Organizing Universe: Scientific and Human Implications of the Emerging Paradigm of Evolution*. Oxford, England: Pergamon Press.

Jantsch, E., and C. H. Waddington. 1976. *Evolution and Consciousness: Human Systems in Transition*. Reading, Mass.: Addison-Wesley.

Kasting, J., O. Toon, and J. Pollack. 1988. "How Climate Evolved on Terrestrial Planets." *Scientific American*, Feb. 1988, p. 90.

Kerr, R. A. 1988. "No Longer Willful, Gaia Becomes Respectable," *Science,* April.
Kirk, G. S., J. E. Raven, and M. Schofield. 1983. *The Pre-Socratic Philosophers.* Cambridge, England: Cambridge University Press.
Kitto, H.D.F. 1978. *Greek Tragedy: A Literary Study.* London: Methuen.
Kitto, H.D.F. 1979. *The Greeks.* New York: Penguin.
Koestler, A. 1978. *Janus: A Summing Up.* London: Pan Books.
Lapo, A. V. 1982. *Traces of Bygone Biospheres.* Moscow: Mir Publishers.
Lappe, Francis Moore, and Joseph Collins. 1977. *World Hunger: Ten Myths.* San Francisco: Free Spirit Press, Institute for Food and Development Policy.
Lazcano, A. 1986. "Prebiotic Evolution and the Origin of Cells." In Origin of Life and Evolution of Cells. Edited by L. Margulis, R. Guerrero, and A. Lazcano. *Treballs de la Societat Catalana de Biologia,* vol. 39.
Lazlo, Ervin. 1978. *The Inner Limits of Mankind.* Oxford, England: Pergamon Press.
Lilly, John C. 1975. *Lilly on Dolphins.* Garden City, N.Y.: Doubleday/Anchor.
Lovelock, J. E. 1972. "Gaia As Seen through the Atmosphere." *Atmospheric Environment* 6 (579).
Lovelock, J. E. 1982. *Gaia: A New Look at Life on Earth.* Oxford, England: Oxford University Press.
Lovelock, J. E. 1986. "Geophysiology: The Science of Gaia." *The New Scientist.* London, n.d.
Lovelock, J. E. 1988. *The Ages of Gaia: A Biography of Our Living Earth.* New York: Norton.
Lovelock, J. E., and L. Margulis. 1984. "Gaia and Geognosy." In M. B. Rambler. *Global Ecology: Towards a Science of the Biosphere.* London: Jones and Bartlett.
Margulis, L. 1981. *Symbiosis in Cell Evolution.* San Francisco: Freeman.
Margulis, L. 1982. *Early Life.* Boston: Science Books International.
Margulis, L., and R. Guerrero. 1986. "Not <origins of life> but <evolution in microbes>." In Origin of Life and Evolution of Cells, *Treballs de la Societat Catalana de Biologia,* vol. 39. Edited by L. Margulis, R. Guerrero, and A. Lazcano.
Margulis, L., and D. Sagan. 1986. *Origins of Sex: Three Billion Years of Genetic Recombination.* New Haven, Conn.: Yale University Press.
Margulis, L., and D. Sagan. 1987. *Microcosmos: Four Billion Years of Evolution from Our Microbial Ancestors.* London: Allen & Unwin.
Maturana, Humberto R., and Francisco J. Varela. 1987. *The Tree of Knowledge: The Biological Roots of Human Understanding.* Boston: Shambala.

Mellaart, James. 1975. *The Neolithic of the Near East.* New York: Scribner's.
Mellaart, James. 1967. *Catal Huyuk.* New York: McGraw-Hill.
Merchant, Carolyn. 1980. *The Death of Nature: Women, Ecology and the Scientific Revolution.* San Francisco: Harper & Row.
Newman, James R., ed. 1956. *The World of Mathematics,* 4 vols. New York: Simon & Schuster.
Ong, W. 1982. *Orality and Literacy: The Technologizing of the Word.* London: Methuen.
Pankow, Walter. 1979. "Openness as Self-Transcendence." In E. Jantsch and C. H. Waddington, eds. *Evolution and Consciousness.* Reading, Mass.: Addison-Wesley.
Pearce, F. 1988. "Gaia: A Revolution Comes of Age," *New Scientist,* 17 March.
Plato. Reprint, 1956. *Great Dialogues of Plato.* Translated by Philip G. Rouse. New York: New American Library.
Postgate, J. 1988. "Gaia Gets Too Big for Her Boots," *New Scientist,* 7 April.
Prigogine, Ilya, and Isabelle Stengers. 1984. *Order Out of Chaos: Man's New Dialogue with Nature.* New York: Bantam.
Rothstein, Jerome. 1985. "On the Scientific Validity and Utility of the Living Earth Concept and Its Further Generalization." Symposium, "Is the Earth a Living Organism?" Amherst, Mass.: University of Massachusetts.
Russell, Bertrand. 1961. *History of Western Philosophy.* London: Allen & Unwin.
Russell, D. A. 1979. "The Enigma of the Extinction of the Dinosaurs." *Annual Review of Earth Planetary Science* 7:163–82.
Sagan, D., and Margulis, L. 1983. "The Gaian Perspective of Ecology." *The Ecologist* 13(5).
Scientific American. 1983. "The Dynamic Earth." September.
Skinner, B. F. 1975. *Beyond Freedom and Dignity.* New York: Bantam.
Sonea, S., and M. Panisset. 1983. *A New Bacteriology.* Boston: Jones & Bartlett.
Spretnak, Charlene. 1981. *Lost Goddesses of Ancient Greece.* Boston: Beacon Press.
Stone, Merlin. 1976. *When God Was a Woman.* New York: Harcourt Brace Jovanovich.
Swimme, Brian. 1984. *The Universe Is a Green Dragon.* Santa Fe: Bear.
Teilhard de Chardin, Pierre. Reprint, 1959. *The Phenomenon of Man.* London: William Collins.
Thomas, Lewis. 1975. *The Lives of a Cell: Notes of a Biology Watcher.* New York: Bantam.

Toffler, Alvin. 1980. *The Third Wave*. London: William Collins.

Toynbee, Arnold. 1972. *A Study of History*. Oxford, England: Oxford University Press.

Varela, F. J., H. R. Maturana, and R. Uribe. 1974. "Autopoiesis: The Organization of Living Systems, Its Characterization and a Model." *Biosystems* 5:187–96.

Vrooman, J. R. 1970. *René Descartes*. New York: Putnam.

Wallace, R. A., J. L. King, and G. P. Sanders. 1984. *Biosphere: The Realm of Life*. London: Scott Foresman.

Watson, Lyall. 1979. *Lifetide: A Biology of the Unconscious*. London: Hodder & Stoughton.

Whitehead, A. N. 1926. *Science and the Modern World*. New York: Macmillan.

Whitehead, A. N. 1979. *Process and Reality*. New York: Free Press.

Whorf, Benjamin Lee. 1956. *Language, Thought and Reality: Selected Writings*. Edited by John B. Carroll. Cambridge, Mass.: M.I.T. Press.

Wilson, E. O. 1975. *Sociobiology*. Cambridge, Mass.: Harvard University Press.

Xenophon. Reprint, 1979. *Recollections of Socrates and Socrates' Defense before the Jury*. Translated by Anna J. Benjamin. Indianapolis: Bobbs-Merrill.